Applied Computing

Springer

London
Berlin
Heidelberg
New York
Barcelona
Budapest
Hong Kong
Milan
Paris
Santa Clara
Singapore
Tokyo

The Springer-Verlag Series on Applied Computing is an advanced series of innovative textbooks that span the full range of topics in applied computing technology.

Books in the series provide a grounding in theoretical concepts in computer science alongside real-world examples of how those concepts can be applied in the development of effective computer systems.

The series should be essential reading for advanced undergraduate and postgraduate students in computing and information systems.

Books in the series are contributed by international specialist researchers and educators in applied computing who will draw together the full range of issues in their specialist area into one concise authoritative textbook.

Forthcoming titles in this series include:

Linda Macaulay
Requirements Engineering

Deryn Graham and Tony Barrett
Knowledge Based Image Processing

Jan Noyes and Chris Baber
Designing Systems

Derek Wills and Rob Macredie
Applied Computer Graphics

Derrick Morris, Gareth Evans, Peter Green
and Colin Theaker

Object Oriented Computer Systems Engineering

 Springer

Derrick Morris, BSc, PhD
David Evans, BSc, PhD
Peter Green, BSc, MSc, PhD
Department of Computation, UMIST, PO Box 88, Manchester M60 1QD, UK

Colin Theaker, BSc, MSc, PhD, CEng, FBCS, AMIEE
School of Computing, Staffordshire University, PO Box 334, Beaconside,
Stafford ST18 0DG, UK

Series Editors

Professor Peter J. Thomas, BA, PhD, AIMgt, MIInfSc, MBCS, MIEEE, MIEE,
CEng, FRSA, FVRS
Centre for Personal Information Management, University of the West of
England, Coldharbour Lane, Bristol, BS16 1QY, UK

Professor Ray J. Paul, BSc, MSc, PhD
Department of Computer Science and Information Systems at St. John's,
Brunel University, Kingston Lane, Uxbridge, Middlesex UB8 3PH, UK

ISBN 3-540-76020-2 Springer-Verlag Berlin Heidelberg New York

British Library Cataloguing in Publication Data
Object oriented computer systems engineering. - (Applied computing)
   1.Object-oriented programming (Computer science)
   I.Morris, Derrick
   005.1'1
   ISBN 3540760202

Library of Congress Cataloging-in-Publication Data
A catalog record for this book is available from the Library of Congress.

Typesetting: Editburo, Lewes, East Sussex
Printed and bound at the Athenæum Press Ltd., Gateshead, Tyne and Wear
34/3830-543210 Printed on acid-free paper

Preface

This book addresses issues concerning the engineering of system products that make use of computing technology. These systems may be products in their own right, for example a computer, or they may be the computerised control systems inside larger products, such as factory automation systems, transportation systems and vehicles, and personal appliances such as portable telephones.

In using the term engineering the authors have in mind a development process that operates in an integrated sequence of steps, employing defined techniques that have some scientific basis. Furthermore we expect the operation of the stages to be subject to controls and standards that result in a product fit for its intended purpose, both in the hands of its users and as a business venture. Thus the process must take account of a wide range of requirements relating to function, cost, size, reliability and so on.

It is more difficult to define the meaning of computing technology. These days this involves much more than computers and software. For example, many tasks that might be performed by software running in a general purpose computer can also be performed directly by the basic technology used to construct a computer, namely digital hardware. However, hardware need not always be digital; we live in an analogue world, hence analogue signals appear on the boundaries of our systems and it can sometimes be advantageous to allow them to penetrate further.

Thus the rather clumsy title of microelectronic and software technology more accurately identifies the technology required in modern computer system implementations. Several properties of microelectronic technology make it attractive for this role. Clearly, where the system products need to be portable, they need to be small, but they also need to be economical in their use of electrical power. Also the more ambitious products will have very substantial complexity, demanding the vast numbers of logic components that can only be assembled by using the automation of microelectronic technology. Of course, software will normally be the means by which the higher level, less time critical functionality is realised. If an integrated approach is to apply to computer system development, which can make optimum use of both microelectronic and software components, it must treat these components in similar ways until a design has been completed and evaluated.

The main objective of the book is to propose a new approach to the development of computer system-based products of the kind we have identified. It aims to meet the engineering requirements of such a process by specifying an integrated sequence of development phases that are supported by notations, techniques, methods and tools. Too few of these approaches exist.

The authors have planned the book to be attractive to both the serious student of systems engineering and practitioners involved in the development of these systems. Its use is supported by tools that are readily available via the World Wide Web, and the proposed methods have a strong focus towards the application of the notation and techniques to industrial-strength problems. We believe the approach as a whole to be unique, but we have drawn freely on earlier work in the areas of Structured and Object Oriented Methods. Practical experience has also formed a strong input and, to this extent, we are grateful to those in industry who have given support and guidance to our research.

It is appropriate to acknowledge that, although four authors have produced this book, many present and former colleagues and students have made invaluable contributions to the evolution of the ideas, techniques and models contained within it. The following have made direct contributions to the book: Richard Barker, Martin Bland, Ronnie Beggs, Willie Love, Phil James-Roxby, Paul Rushton and Simon Schofield. Last but by no means least we wish to record our debt to Professor Roger Phillips of the University of Hull, who has been associated with the work from its outset and has only relaxed his direct involvement at this stage because of the pressure of being a Head of Department.

Special mention must also be made of the support that our work has received from funding bodies. The Science and Engineering Research Council (SERC), now known as the Engineering and Physical Sciences Research Council (EPSRC), has funded research leading to the development of the ideas presented herein, in contracts GR/J07631, GR/J09505, GR/J09840, GR/K60923 and GR/K61036. The Open Microprocessor Initiative (OMI) within the European Community ESPRIT Programme has supported the evaluation of tools and methods that apply the ideas behind the approach, through contracts OMI/DE 6909 and OMI/MODES 20.592, and perhaps even more importantly the Initiative has provided a framework which has brought those working on the project into contact with the leading edge of microprocessor development in Europe.

Contents

1 Introduction to Computer Systems

Computer systems are everywhere. One has only to walk into any shop, office, station or airport in the industrialised world to be confronted by a host of systems handling an enormous range of applications, from those concerned with information storage, retrieval and manipulation, to those in which the computer system plays a more pro-active role in work by providing intelligent support for management, design and production tasks. In addition to these easily recognised computer systems, there is an even larger class that is concealed within other products encountered in normal daily life, such as washing machines, microwave ovens, telephones, cars, trains and aircraft. Furthermore, in recent times the potential for an even larger proliferation of computer systems has arisen in the form of computer driven 'gadgets' that we might carry with us wherever we go. Thus, the ubiquitous nature of computer systems requires no further comment.

Considering the impressive achievements of this growing industry, what is perhaps surprising is that the discipline of developing such systems, Computer Systems Engineering, is by no means mature or stable and is only now becoming the focus of a significant amount of attention. There can be little doubt that the development of a computer system is a daunting and challenging task, the complexity of which both fuels the demand for a well-formulated approach to the engineering of these systems and militates against that demand being satisfied. The scale of complexity is almost beyond comprehension. Typically, a large system might contain hundreds of interconnected computers coupled to special-purpose hardware devices and millions of lines of software. Even simple products, such as television sets, can contain up to 500 kilobytes of software. However, these static measures only reveal part of the problem. Another hint of the difficulties of fully understanding what is happening inside a computer system is found in the rate at which modern computers can obey their instructions, typically 50 million instructions per second (MIPS) for higher end processors.

Although significant progress has been made with techniques for engineering the hardware and software components of a computer system, there is relatively little that applies to the design of the system as a whole. Even research has tended to target specific issues, such as techniques of abstraction, formalisation of the specification of behaviour and structure, and languages to support component synthesis through compilation. Less attention has been paid to methods for engineering whole systems, which provide an integrated approach to both hardware and software elements. Indeed, one popular view of a computer system that reflects the lack of a total system culture is that *functionality is largely provided by its software, and hardware is selected to provide an appropriate execution platform for the software.* Another view, common for example in signal processing applications, is that *the hardware has to be designed first in order to deal with time-critical aspects of the process, and software is added to assist with the more tricky aspects.*

In contrast, this book presents an integrated approach to the design of the hardware and software of a system. The integration is provided by a lifecycle that starts with **cospecification** to produce a model identifying the behavioural requirements of a complete product. This **behavioural model** provides the focus for the integrated approach as it undergoes a series of enhancements and transformations. In the middle stages, the model is made executable in order to provide a **cosimulation** environment in which system-wide issues can be addressed prior to the detailed design and construction of system components. Decisions concerning the relative roles of hardware and software are evaluated by experimentation with the executable model, and they are introduced into the model by a technique known as **transformational codesign**. The philosophy underlying this approach recognises that, while software may control overall functionality, the performance, size and power requirements of a system are largely determined by its hardware, and that some system objectives only become feasible if the design process allows functionality to be transferred between software and hardware. The 'best' design will then be the one that achieves an optimal implementation of functionality. To allow this flexibility, behavioural models have functional components that are uncommitted with respect to implementation technology, and it then follows that compatible products within a range may be developed from the same model.

1.1 STRUCTURE OF THE BOOK

The remainder of this chapter defines the kinds of computer system for which the techniques presented in this book are intended, summarises the technologies that are used to manufacture the systems, provides an introduction to computer systems engineering, identifies the characteristics of these systems that impact the problems of engineering them, and finally, illustrates the potential pitfalls with anecdotal evidence of major problems that have arisen in various ambitious computer system developments.

Chapter 2 is a state-of-the-art survey of approaches used to engineer computer systems, and Chapter 3 complements this 'process overview' with a survey of techniques for analysis and design that are commonly used within the process of developing systems. At the end of Chapter 2 there is an introduction to the integrated approach to computer system development proposed by the authors. This approach to System Engineering is both Model-based and Object Oriented, hence it is known by the acronym MOOSE (Model-based Object Oriented Systems Engineering).

Chapters 4 to 8 deal with all the phases of integrated product development, focusing particularly on the MOOSE approach. The application of the techniques is illustrated within these chapters through the development of a typical computer system product, a video cassette recorder (VCR). Tools to support computer systems engineering naturally feature throughout and there is particular reference to a MOOSE workbench that is freely available via the World Wide Web. The success of any approach may depend on practicalities as much as on abstract elegance, and so Chapter 9 discusses the pragmatics of applying MOOSE, particularly in so far as it affects significant product developments. Further supporting detail on MOOSE, and other design examples, will be found in the Appendices.

1.2 DEFINITION OF COMPUTER SYSTEMS

The computer systems towards which the techniques described in this book are directed are:

> Systems that consist of one or more processors, together with associated software and hardware (which may be both analogue and digital). These systems often need to communicate with their environments in application specific ways.

There are a number of important implications in this definition.

First, the term 'processor' is used rather than 'computer' because it implies the most general form of information processing. A processor is *'a machine or system which performs a process'*, while a computer is *'a calculating machine'* (Simpson and Weiner 1989). Second, we are concerned with complete systems, which combine both hardware and software and an integrated approach to their development, rather than concentrating separately on the engineering of the hardware or software components from which systems are created. An enormous amount of research and development effort has gone into devising approaches to engineer system components but, as we shall discuss later, engineering systems that are built from many such components is still a hazardous business in spite of this work. Third, the definition acknowledges, albeit implicitly, that for computer systems to be useful they must interface in a suitable way with the real world, and this may require the custom building of both the hardware and software that provides the interface.

The definition given above is intentionally very broad, as computer systems have very wide application. However, the methods introduced in this book are concerned more with computer system-based products than with what might be termed 'Information Management Systems'. Systems of particular interest therefore include reactive systems, defined by Harel (1987) as:

> ...characterised by being, to a large extent, event driven, continuous-ly having to react to external and internal stimuli. Examples include telephones, automobiles, computer operating systems, missile and avionics system, and the man-machine interface of many kinds of ordinary software.

They also include embedded systems, defined by Zave (1982) to be '...embedded in larger systems whose primary purpose is not computa-tion...', and which are used for '...process control: providing continual feedback to an unintelligent environment...'

Such systems have wide application and vary greatly in scale from high performance, multiprocessor systems used in industrial process control, aerospace, etc. through to single processor systems in personal communications systems and home appliances (video cassette recorders, microwave ovens, etc.).

The broad class of systems termed information management systems have different characteristics stemming from a large quantity of inter-nally stored information and intrinsically (relatively) simple processes that provide for user access and manipulation of the stored information. However, trends towards the large scale distributed information systems and their exploitation of hardware dependent interfaces, such as multi-media, are creating a need for personalised embedded systems within the information management system sector.

Common examples of information processing systems are payroll sys-tems, ordering and invoicing systems and flight reservation systems. These kinds of system are often characterised by having well defined input, output and user interactions and they consist of software running on commercially available hardware platforms. This is not to say that the engineering of such systems is easy, or that the approach discussed in this book is unsuited to the development of these systems. However, the authors acknowledge that this type of system is often information inten-sive rather than process intensive. Thus, an information modelling rather than a behavioural modelling approach may be required.

1.3 COMPUTER SYSTEMS TECHNOLOGY

At a very superficial level we might say that computer systems technology is the aggregation of software technology and hardware technology. Indeed, many systems have been built by splitting the task into two and producing the hardware and software by totally separate development efforts. The integrated approach to computer system development explored in this book requires a more subtle blending of the two technologies and the addition of quite a lot that is new. However, the two base technologies are still present and it is appropriate to summarise them.

Software is, without exception, prepared as a source text in one of a large number of programming languages. Computer systems engineers will normally elect to use a language such as the assembly language specific to the processors in the system, a higher-level language such as C, an object oriented language such as C++, or a real-time programming language such as Ada. Analysis and design tools may be used before the code is cut, although experience and opinions in this dimension show wide variation. Finally compilers, debuggers and development environments help the engineer to produce and test the binary code for the system.

For hardware a wider variety of technologies is available. At the 'high tech' end of the spectrum, where the hardware is to be manufactured in a silicon foundry, quite complex systems can be manufactured automatically from a VHDL (VHSIC Hardware Description Language) specification as a single chip (*deeply embedded system*). For example, high performance 32-bit processors are routinely embedded in single-chip systems to provide high levels of system functionality, and the potential for combining standard processor cells, special purpose processors, hardware functions and software on the one chip is rapidly increasing. If the foundry approach is not appropriate – for example, because the sales volume of a product does not justify it – Printed Circuit Board (PCB) technology is used to combine the components. However, there may still be large areas of functionality covered by custom hardware implementations in the form of Application Specific Integrated Circuits (ASICs).

At the other end of the technology spectrum the hardware of a computer system might consist of a network of processors running a distributed application. Somewhere between the two levels, the hardware might be standard workstations enhanced by mounting application specific boards on the processor bus.

In spite of the wide variation in hardware technology and significant variations in product volume, a unified approach to computer system development is nevertheless desirable because the products still have to be engineered to a competitive cost and to meet 'time-to-market' deadlines. In contrast, traditional practices have grown out of the separation of hardware and software development, and experience has shown that such an approach is fundamentally flawed. For example, a typical approach to producing an embedded system product might be:

- First select a standard microprocessor based more on historical practice than sound evaluation of its suitability for the product under development.
- Then develop a 'bread-boarded' version of the projected hardware on which software development can take place.
- After this, hardware design and development will proceed independently of software production, and the hardware effort produces an ASIC from which an Alpha system can be constructed.
- Next, the process of integrating software with hardware can begin. This is usually a painful process, as a result of which there are usually repercussions back into the design of both the hardware and software.
- A Beta system will follow after some re-engineering, but...

Already in the above approach, three systems have been built. The result is probably not optimum, the timescale is protracted, maintenance of the product is likely to present difficulties and, in severe cases, the product may not meet the requirements. Such experiences establish the case for the integrated approach presented in this book, where we concern ourselves with designing the *system* before we consider the technology needed to implement the components of the system. Thus the majority of the book focuses upon the development of the system rather than the implementation of its components. Of course, the physical realisation of the system in one of the many hardware technologies available is extremely important to the success of the product. We demonstrate this in Chapters 8 and 9, and show how the approach taken is flexible enough to develop hardware for a wide range of viable implementation technologies.

1.4 INTRODUCTION TO COMPUTER SYSTEMS ENGINEERING

This section presents a condensed view of the process of engineering a computer system, which will serve to introduce the general principles. Chapter 2 expands on the topic and presents a more detailed discussion of the steps that make up the process. However, regardless of the way in which the process is broken down, there are three essential stages involved in the development of a successful product, namely:

- Capture and approval of a satisfactory set of system requirements.
- Generation of a design that satisfies these requirements.
- Realisation of an implementation that conforms to this design.

Any proposed development process should be evaluated on the basis of how well it deals with these three aspects.

The development process must also perform well with respect to two other critical factors which greatly affect the commercial success of a product. The first of these is that budget and time constraints on the development have to be met. Obviously an overpriced product will not sell in a competitive market, and for many product opportunities the time-to-market is just as crucial. Although development times are difficult to predict, experience has shown that the market does not stand still and the potential volume of sales is soon eroded by competition. Hence the emphasis has to be on structuring the development process to fit the time-to-market margin rather than on prediction of the time needed. This time pressure inevitably fuels a demand for techniques that allow developments to be based on accumulated experience. The development process therefore needs to be geared seriously to allow for re-use of both the design and implementation of components, since this is the only way of achieving a competitive time-to-market.

The other critical factor that impinges on the success of a development process concerns the quality of the documentation that it produces. Quality in this sense means the contribution that the documentation makes to the smooth working of the process. For example, the ease with which the design and implementation documentation for a system can be understood by all those with a role to play in the process has a significant impact on the productivity and product quality. It also has longer term impact, for example, on carrying the 'know-how' through to future generations of product engineers and supporting the vital re-use of design and implementation achievements.

1.4.1 Operation of a Computer Systems Engineering Process

Computer system developments usually require teams of design and implementation specialists led by experienced product engineers, possibly interacting with product planners. These teams operate inside a management framework concerned with the organisation's business interests and strategies. Thus, before embarking upon further discussion of computer systems engineering, it is useful to postulate a typical organisational structure in which it might be applied. Of course, one typical organisation will not cover all situations, but a specific example is needed to set a context in which the role of methods, techniques and tools can be clarified.

In so far as it impacts on methods and techniques, it will be assumed throughout this book that, when a new computer system product development is started, a senior engineer, hereafter referred to as the Project Engineer, is assigned to lead the development team. This individual will have overall technical responsibility for the project, and in particular will control the production of deliverables that must emerge from a sequence of defined points in the development process, and steer those deliverables through any evaluation and assessment reviews that the process imposes. However, for a product to be successful, there is still a need for intuitive skills and imaginative talent, and the Project Engineer must have these, together with the freedom to apply them in making critical systems engineering judgements as and when necessary. Hence the process is defined more by its sequence of deliverables than by formulae for producing them.

In many methods the deliverables take the form of 'models' that act as the vehicles to support reviews and assessments, and they also carry firm, well documented decisions from one stage to the next. In their review role, models provide a mechanism for attracting the input of independent judgements. A strongly model-based approach will also use models to support experimentation in areas where uncertainty arises, to facilitate training of the personnel to be involved in the post-delivery life of a product, and to stimulate re-use by carrying know-how and detail from one product development to another.

Typically, the models used in a computer system development process will have a hierarchical structure which provides a framework which enables the Project Engineer to allocate responsibility for detailed development of subsystems to specialist design teams, while still maintaining enough design control to retain responsibility for the end result. A very large system may require several project teams, possibly in different companies, and methods should extrapolate to this scenario, with the models facilitating collaboration and agreement across team boundaries.

1.4.2 Phases in a Computer System Development Process

Figure 1.1 provides a very simplified view of a hypothetical process that might be used for the development of computer system products. In practice, the process would also be iterative, as problems identified in the later stages of development may require the earlier stages to be re-visited.

The Analysis of Requirements phase produces a Product Specification, which must be sufficient to ensure that the right product will be built. Clearly, the specification must define the required functional behaviour of the product, together with any other non-functional requirements of the product that are considered important to its commercial success; for example, size and cost limits. How else would the engineers know the criteria for selecting between functionally equivalent alternatives?

The form of the input into the analysis phase, and hence the type of work it involves, can be highly variable. For example, a detailed textual Requirements Specification may be produced prior to the start of the development, which is the result of an exhaustive study of the market and of the potential users. This Requirement Specification could give a description of every facet of the required system's behaviour and a detailed analysis of the constraints it must satisfy. For a more speculative

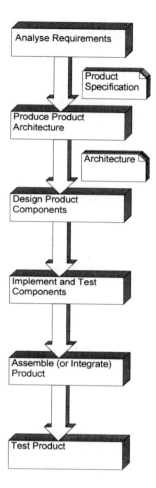

FIG. 1.1 A hypothetical computer system development process.

development, the process might start from a plain statement of a desire to develop a particular type of product at a specified cost and within a tightly constrained timescale. In this case the development team will need to apply imagination and invention in order to prepare a detailed Product Specification for approval. The imaginative aspect is particularly important for innovative computer system product because, while the end product must fit user needs, there will be no current user community from whom to solicit opinions.

Ideally, because of its role as a form of contract between the system development team and their 'customers' (who may be representatives of an external agency, an internal marketing department, senior management, etc.), the Product Specification must use language understood by all and be free from ambiguity.

Following on from the analysis phase, Fig. 1.1 allocates two phases for design type activities, the first of which acknowledges that, before embarking on component design, a system architecture is needed in which the functional role of each subsystem is made clear and any constraints on the operation of the subsystems are identified. Obviously, the total system represented by the architecture must provide the functionality identified in the Product Specification, satisfy the non-functional requirements, and make clear the interplay between the components. It must also provide a clear identification of the contribution of each component to the overall requirements. Hopefully the allocation of appropriate non-functional requirements to individual components or subsystems, or alternatively the imposition of constraints on the design of these parts, will result in overall compliance with the non-functional requirements of the product.

Definition of an architecture is a task for the Project Engineer and possibly a few senior colleagues. It is a bottleneck in the process whose duration cannot be reduced by applying more effort. However, once established, the architecture opens the door for the rest of the development team to work in parallel on developing component designs.

In many current approaches to computer systems engineering, an architecture would be evolved by intuitive techniques, and the design and implementation of hardware and software components would then proceed concurrently – and hence independently of each other. This informal approach has some appeal, although it opens up many possibilities for serious errors. For example, not too many hard decisions have to be faced 'up front'. In order to reduce the risks due to errors, when time permits, some prototyping experiments and feasibility studies may be carried out before finalising the architecture. Again, such decisions are often taken informally. It will be seen later that the approach on which this book is based imposes tighter control on risk analysis and risk elimination.

Summarising the above, in principle the Product Specification defines *what* product is to be built and the designers decide *how* the specified functional behaviour is to be provided. Their choice should take account of the non-functional requirements and produce optimum solutions. Design is a painstaking activity that needs to be carried out with precision, although the 'back of an envelope' syndrome is hard to resist. Some methods of design and the notations employed are discussed in Chapter 3.

It might seem that the next phase, that of generating implementations from the designs, should be straightforward, but very often this is not so. One problem is that the boundary between design and implementation is not well defined. Hence, even if the designers have been diligent, the implementor might have to interpret unclear and sometimes even ambiguous designs in order to 'guess' the intent. This can be a serious source of errors. Indeed there is much scope for error in producing implementations, and the techniques for achieving the transition from design to implementation are an important issue to be resolved in planning a development process. Again in Chapter 3 some current techniques for moving from design to implementation are summarised.

Ultimately a development process has to recognise that there is the possibility of errors at all stages. Even with good technique, errors will occur, and so implementations have to be tested and bugs have to be fixed, as shown in Fig. 1.1. This testing and correcting of errors tends to occupy an embarrassingly large period of time at a stage when there is usually not much time-to-market remaining.

1.5 THE CHARACTERISTICS OF COMPUTER SYSTEMS

All computer systems are developed to satisfy a set of functional requirements. In addition, the design of the computer systems of particular interest here is always strongly constrained by requirements that are not functional. For example, in portable devices, their size, weight and battery life might be important factors. In general, functional requirements derive from an analysis of the purpose of the product, whereas non-functional requirements originate from the environment and context in which it is to be manufactured, marketed and used.

Since, to many of us, the computer systems concealed in other products are less visible than those that pay our salaries and send us bills, their importance to national economies might not be recognised. However, embedded systems have, for many years, been playing an important economic role. As early as 1978, the US Department of Defense was spending 56 per cent of its $3 billion annual software budget on embedded systems (Zave 1982). More recently (Zave 1994) it was estimated that there were around 140 million personal computers (PCs) in the world, but three times as many processors embedded in other systems. It is clear that developments in silicon technology and microprocessor design have led to embedded systems becoming a basic enabling technology for the whole of the manufacturing industry, from white goods to aerospace (European Commission 1991). It is in markets of this kind that overpriced or late-to-market products are certain to be commercial failures.

1.5.1 Functional Characteristics

The functionality, or behaviour, of a system refers to what the system does and is not concerned with how often or how efficiently it does it. Computer systems have, by definition, substantial software components and, as such, are capable of exhibiting highly sophisticated functionality. By making use of application specific hardware they can also provide enhanced performance for some of this functionality. This is indeed one of the reasons why computer systems are increasingly being embedded in other systems for control purposes.

The behaviour of a computer system can be specified purely in terms of the required responses to a set of input stimuli. A system's response to a stimulus may be dependent on the receipt of other stimuli and on the internal state of the system. Products with complex functionality generally have a large set of stimuli and responses, and their interdependencies imply a large number of different internal states. Adequately specifying and designing the behaviour of the system for all combinations of stimuli and state is perhaps the largest complication in their development, and Brooks (1987) differentiates between engineering the software for computer systems and the engineering of other kinds of systems on this basis.

Thus computer system development methods must be able to capture and represent the functional attributes of the system and support the analysis that allows developers and their customers to agree on complex system functionality. The method must allow the agreed functionality to be carried through into design and implementation and be audited and validated at stages along the way. Problems concerning the partitioning of a system into hardware and software, and distributing its actions and control over a specific set of components, increase in importance as the functional complexity of systems increases. Hence models to capture and clarify functional behaviour are needed. The impact of design decisions on the non-functional characteristics of the product can then be investigated in the context of these functional models.

1.5.2 Non-functional Characteristics

The significant non-functional attributes that must be satisfied by reactive computer systems include performance and timing, reliability and safety criticality, concurrency and cost, and these will now briefly be reviewed.

Performance and Timing

Performance is a term that is often used to refer to a wide range of concerns in the context of computer systems. For instance Kant (1992) uses this term to encompass reliability and productivity, which in his definition includes user interface requirements. Here the term performance is used with a more limited scope, in a way that is consistent with its use in other engineering systems. For example, a car's performance is specified by a set of maxima and minima (maximum speed, maximum acceleration, minimum fuel consumption, etc.) derived under ideal conditions. Hence, with respect to computer systems, performance is taken to refer to parameters such as response time, throughput, etc.

We also distinguish here between performance and timing, as we believe they represent different perspectives of the temporal behaviour of a system. For example, both may specify the response times of the system to a specified input stimulus. However, performance pertains largely to the mean response times of the system, and the design goal is generally to minimise these times while maximising the resource utilisation and throughput. In contrast, timing behaviour is concerned with meeting specified deadlines, and in particular the predictability and temporal correctness of the system. Consequently resource utilisation is of secondary importance.

System performance is an important issue. As a general rule, performance can always be increased if one is prepared to increase the system cost. Therefore the performance goals of a system should be identified in advance and the most cost-effective way of satisfying them should be selected.

A computer system development method must support the capture of performance attributes and the analysis of the performance of a design. This is generally achieved by simulation modelling, analytical calculation or by experimenting with similar systems which already exist.

In terms of timing, most embedded computer systems have stringent constraints placed upon their operation. These are derived from their operating environment and identify the deadlines by which the system has to respond to stimuli. The consequences of missing a deadline may vary, and so systems may be classified as either soft real-time systems, where the timing requirement should be met but the consequences of occasionally missing a deadline are not serious, or as hard real-time systems, where a specific subset of the timing requirements must always be met if serious consequences are to be avoided. If a critical deadline is missed in a hard real-time system, the system is deemed to have failed.

There are two distinct yet related attributes to be considered within timing constraints, namely responsiveness and precision. The first is that, for a given stimulus, the response must be derived within a given time. For example, a car's engine management system will determine when a sparkplug should fire, based upon the sampling of a number of inputs. The computation has a real-time deadline since it must be complete before the firing time when the piston approaches top dead centre. The second attribute, precision, is concerned with the computer system carrying out actions at precise instants in time – not before or after a deadline, but exactly on it. When the engine management system has determined when the sparkplug should fire, the system must cause the plug to fire at that time (although there will be a small tolerance around the actual optimum instant of firing).

Timing attributes must be captured by a computer system development method, and it should be shown that the design and implementation meet these requirements. In practice, showing that a software based system will meet timing requirements under all circumstances is far from easy (Stankovic 1988). The use of formal language notations has done little to help in this area, as these tend to focus on temporal correctness rather than on the quantifiable behaviour of the system. Although correctness is important, it still leaves a measure of uncertainty in the product design.

Reliability and Safety Criticality

The reliability of a system, defined by Burns and Wellings (1989), is 'a measure of the success with which a system conforms to some authoritative specification of its behaviour. When a system deviates from this specification it is said to fail.'

The reliability of any product is critical to the success of the producer; a reputation for unreliability has caused many technically innovative products to fail commercially. While product reliability is usually critical to the commercial success of the producing company, in many applications the reliability of the product is also critical to the economic strength or health of the product's users, customers and society in general, because computer systems are being increasingly used in safety critical applications such as aircraft, power generation and chemical plant control.

Systems fail, broadly speaking, for one of five reasons:

- Inadequate specification.
- Design errors.
- Implementation errors.
- Component failure.
- Interference.

The first of these should be addressed by using appropriate system development methods such as those discussed in Chapter 2, although these offer no guarantee of success or, perhaps more significantly, no quantifiable, statistically valid prediction of system reliability.

Design errors should also be addressed by development methods. However, software in particular is difficult to analyse due to its high number of internal states, and exhaustive testing is generally not possible. This is also becoming true of complex hardware systems, such as the current generation of high performance processors. Reliability is generally addressed by incorporating redundancy in the design, such as N-version programming, recovery blocks, etc. (Littlewood and Miller 1991).

Implementation errors are concerned with defects in the manufacturing process. For example, these may be due to faulty welding, marginal wafer fabrication, poor soldering and programming errors. Quality engineering techniques may be used partially to address these problems for physical systems. For software, the problems are difficult to detect and require safe implementation procedures and extensive testing.

Component failure is primarily limited to hardware, since software does not wear out. Also, as hardware faults are statistically independent and reliability testing is well understood, obtaining statistically valid estimates for the reliability of hardware subsystems is routine through reliability modelling.

Finally, interference is best handled by the use of design techniques either at the hardware level (fault removal) or at the system level using data diversity or N-modular redundancy (fault tolerance).

A computer system development method should allow an assessment of a system's reliability. However, for software based systems, and thus by extension for computer systems, this is far from easy.

Concurrency

Reactive systems are typically required simultaneously to monitor a number of incoming data and event sources and to control a number of external units. This requires the concurrent operation of a number of 'subsystems'. Hardware naturally supports concurrency, but software is naturally sequential. To support concurrent operation, software process-es must either be assigned a single processor each (the system will then exhibit true concurrency) or they must be multi-tasked to use a single processor's thread of control (the system will then exhibit pseudo-concurrency). The extent of the true and the pseudo-concurrency then has implications in terms of a system's performance, cost, size, power requirements, etc., and these are often interrelated.

Methods to support computer system design should, at the analysis stage, expose the full range of concurrent operations to the system devel-oper, as generally it is more difficult to introduce in the later stages of development. During the design phase the method must allow the designer to reduce the concurrency to levels that satisfy the system's requirements, particularly performance and time-critical deadlines. In addition the concurrent operation of subsystems requires that synchro-nisation and interprocess communication mechanisms be provided.

Cost

In the most general terms, the cost requirements of a product can be divided into the development cost and the manufacturing cost. The development cost is that incurred by the company in moving from a product idea to the first tested version of a product. The manufacturing cost is the cost to a company of producing the physical product and is made up of raw material cost and overheads such as manufacturing plant and labour costs.

Obviously a company will seek to control and minimise the development costs of a product. Cost models are used and they can be quite accurate for system developments where the developed product is similar to those that have been developed before. For example the COCOMO (Boehm 1981) model has been used with reasonable success for estimating the development costs of software projects. The same technique can be applied to hardware, particularly when the method of producing it follows a well defined pattern and uses well defined notations. However, there is less practical evidence for this assertion than there is in the case of software. In principle, these methods rely on estimating the scale of the development, for example of the number of lines of code to be produced, and this is, of course, difficult when the development is innovative and extends beyond the developing company's experience. Also, although predicting and controlling costs is a step in the right direction, minimising them is the main requirement. In this respect, we must not overlook the significant cost in terms of lost revenue by being late on the market with a new product.

Computer system development methods must therefore seek to support, control and above all minimise development costs. This can be achieved, in part, by providing an assessment of risk in the early stages of the lifecycle, when the problematic and potentially iterative parts of the development process can be identified. The impact of different development strategies on the time-to-market for the product can be assessed, including the opportunities for component reuse. Thus, a broad strategy is to analyse risks and to minimise them before embarking on detailed design.

The manufacturing cost must be considered during the product development and be traded off against other requirements, notably functionality and performance. For software systems, the manufacturing costs are generally very low; for computer systems they are usually a significant part of the product's selling price.

1.6 RECORDED EXPERIENCES WITH COMPUTER SYSTEMS

Research into methods and techniques for engineering computer systems, particularly the software, is as old as the computer business itself. Although significant advances have been made over three decades, the problems of engineering software still remain and are widely reported (see, for example, Brooks 1987), and developing quality software systems on time and within budget remains difficult. Perhaps part of the reason for this is that 'the target moves'; system size and complexity continue to increase and the tolerance level for error reduces, particularly in safety critical systems (Parnas, Schouwen and Kwan 1990). However, to believe that we could now hit the target provided that it does not move is perhaps optimistic.

The problem of developing reactive computer systems appears to be an even more hazardous activity than the development of more general systems. This problem stems partially from their heterogeneous nature, which requires the concurrent design and development of both hardware and software, and because the unique characteristics of reactive systems demand that a broad range of requirements be satisfied. Not only are the developers expected to identify what would be the right system for the purpose, they are also expected to get it right, make it quickly and optimise its non-functional attributes. Much of the problem is concerned with handling complexity at the system level in the analysis and system design phases, and with making the right architectural decisions, rather than in engineering the components that make up the system. It is argued that this is because the latter issues are well understood, although it might also be that the flaws in the process are better recognised.

The following are examples of recently developed systems which have proved to be catastrophic in one way or another. They are by no means exceptional, other than that they are well documented and by virtue of the extent of the loss that has been incurred. In fact, one recent author (Gibbs 1994) presents the extremely bleak picture that, for every six new, large-scale software developments put into operation, a further two are cancelled, software developments overshoot their schedule by a half on average, and three-quarters of all large systems are 'operational failures' that either do not function as intended or are not used at all.

1.6.1 The Denver Airport Baggage System

The scale of the Denver Airport project is impressive by any standards. Following approval in 1989 to start construction, the plan was to have a new airport with five runways operational by the end of 1993. Occupying a site of 53 sq. miles, and with a minimum distance of one mile between parallel runways, the airport could support three simultaneous landings under any weather conditions. A further runway was planned for 1995, with the possibility for future expansion of up to twelve.

The baggage handling system to support the twenty major airlines flying into Denver naturally had to be on a comparable scale (Gibbs 1994). A system was proposed costing $193 million, and was based on 4,000 independent 'telecars' which would carry the baggage across a network of 21 miles of track. Laser scanners were installed to read bar-coded luggage tags and thus control the routing of baggage, and 5,000 photocells tracked the movement of the telecars. The entire system was controlled by a network of some 300 computers.

This ambitious system was beset by problems, mainly attributed to bugs in the software. Telecars were mis-routed or crashed, and baggage damaged or misplaced. Without adequate baggage handling facilities the airport was unable to open; the delay was estimated to cost Denver City and the Airport Authority around $1.1 million per day in interest and other operating costs. After spending a further $88 million to improve the system, the airport opened in February 1995, some fifteen months late. However, the saga continues since, at that stage, only one major airline was using the automated system, the others choosing to use a more conventional system until the automated system is proven.

Many factors contribute to the problems with this type of system, but a critical issue that the developers must address is 'how can the software be tested and validated before it is put in charge of a system of this scale of complexity?'

1.6.2 The Clementine Space Mission

Whereas the previous example illustrated the system problems that can arise due to scale, this example highlights the difficulties with testing, particularly when the system is to operate in a hostile environment.

In January 1994, the US Ballistic Missile Defense Organization (successor to the 'Star Wars' Strategic Defense Initiative) launched the Clementine satellite, with a mission to use its sensors and cameras to survey the surface of the Moon, to conduct various military experiments, and to visit the asteroid belt in a 'close-fly-by' of the asteroid Geographos. Clementine is distinctive in that it is a low cost project (less than $100 million) and uses off-the-shelf advanced technology in its construction. It has therefore been heralded as the start of a new era for space exploration, far removed from the billion dollar programmes that had previously been the norm (Beason 1995).

In its initial stages the mission was an outstanding success. A topographical survey of the Moon was performed from a polar orbit, and over 1.5 million images were recorded. This revealed many features that previous missions had been unable to survey. In May 1994, as Clementine was being readied for the next phase of its mission, the on-board computer system independently activated several of its thruster rockets during a period when telemetry was interrupted between the ground station and the satellite. The effect of the eleven-minute burn on the manoeuvring thrusters was to expend all the fuel in the Attitude Control System tanks. The satellite was then unable to manoeuvre to allow it to visit the asteroid Geographos.

Although much of the technology used in Clementine is of a very advanced nature, the computer system is of a well established technology. The processor is radiation hardened and of an established military standard. The software was also tested to rigorous (and expensive) Department of Defense standards with a view to certifying its reliability. This aspect of the space programme was therefore not experimental. Nevertheless, a computer system fault, believed to be a software bug, had resulted in the mission being prematurely curtailed.

1.6.3 The Intel Pentium™ Processor*

This final example considers the problem with the floating point divide function of Intel's Pentium processor chip. Although apparently less of a 'system' problem than the previous examples, the fault with the early Pentium processors has aptly been described as '*a software bug that's encoded in hardware*' (Halfhill 1995), rather than a hardware defect or error. It also illustrates the problems that can occur when designs are committed to silicon, and the inevitably high cost of remedying problems in mass produced systems.

The background to the problem lies in the desire to implement a fast floating point arithmetic unit, with a performance some three to five times that of an Intel 486 processor with the same clock speed. The 'divide' function is one of the more complex ones to implement and, to achieve the desired performance gain, an algorithm that uses a lookup table is employed. The lookup table has 1,066 relevant entries but, due to a programming error, only 1,061 entries were loaded into the programmable logic array (PLA) of the Pentium's floating point unit. The PLA was not tested to see that the table had been copied correctly.

A consequence of this mistake is that calculations involving a floating point divide, albeit only with certain patterns of numerators and divisors, are likely to produce inaccurate answers. In the worst case, the error can be as high as the fourth significant decimal digit. Opinions are divided on how frequently these errors are likely to occur. For typical spreadsheet users, Intel estimate an error once every 27,000 years, while IBM's estimate is one every twenty-four days. What is incontrovertible is that the processor did not perform to specification as far as the IEEE's (Institute of Electrical and Electronics Engineers) floating point standard is concerned.

This defect has been estimated to have cost Intel upwards of $400 million (Byte 1995).

---

* Pentium is a trademark of the Intel Corporation.

2 Engineering Computer Systems

To produce a complex computer system on time and within budget is a formidable task which requires a systematic and organised approach. It is unfortunate that most current approaches used for developing these systems are far from ideal. Many have been based on the separate development of the software and hardware parts of a system, although some have adapted a Software Engineering method to the engineering of the total computer system. Very few approaches are specifically designed to provide an integrated process for developing a total system.

The main purpose of this chapter is to review the state of the art with respect to the organisation of a Computer Systems Engineering process and the methods it employs, and to introduce a new integrated approach. Also, because of the relatively close fit between Software Engineering methods and the kind of approach needed for computer systems, some sections focus on the engineering of software instead of the total system. In general, existing hardware methods are usually lower-level in nature, are not based on structured approaches, and hence do not offer the same opportunity for extension into a 'total system' approach. Before considering the methods further there is some terminology whose meaning needs clarification.

2.1 TERMINOLOGY OF THE DEVELOPMENT PROCESS

A bi-product of the rapid growth of computing and computer-related disciplines is that those involved use the same words but usually with different connotations. So far in this book, terms like lifecycle, method, technique, notation and tool have been used rather loosely and, in order to make a closer study of the process for engineering computer systems, we need to have a better-defined set of terms.

First we need a term that covers the whole development process, and a definition of the boundaries of this process. In Software Engineering a popular view of product development is that it does not end until the product is no longer supported. That is, the development of a product occurs over its full lifetime 'from cradle to grave'. This view is based on the thought that 'maintenance' of the delivered software will be required, and this must be provided as an integral part of the development process since changes are accomplished by looping back through the earlier stages.

The nature and economics of computer system products are not sympathetic to this protracted view of development. Software maintenance normally covers both product enhancement and the fixing of bugs, but a cleaner cut-off point is required with computer system products. Enhancement is best treated as a new development; for example, producing a mark N+1 version of the product, albeit making significant re-use of the mark N design and implementation. Fixing bugs in delivered computer systems means product recall, and to plan for this as part of the normal development process would be unacceptable. Therefore the development period for computer systems has to be from 'cradle to delivery', which conveniently coincides with the time-to-market margin.

With software systems, the organisational plan for the development process is often termed the *lifecycle*, since it applies to the perceived lifetime of a product. However, this is not the only term used; for instance, the terms *method*, *process model* and *paradigm* also appear in the literature. It is best that we choose one of these terms and use it consistently, although in fact only the term method causes serious confusion because of uncertainty regarding its scope. Other authors have recognised the ambiguity in the term, and some use the term *methodology* instead, presumably because the development process they advocate involves significantly different methods (and techniques and notations) in different phases. However, methodology is also a much abused term. In this book we shall use the term **paradigm** when referring to the whole development process. Furthermore, we shall assume in our case that it applies from the time a product development is commissioned to the time a tested product is ready for delivery.

Typically a paradigm is a model that breaks the development process into a series of phases that deal with different but closely related aspects of the development. In each phase of the paradigm, methods are needed to accomplish the goals of the phase, and techniques and tools are needed to apply the methods. Thus a **method** in this book is defined to be:

A systematic way of proceeding with a well defined phase of development of a computer system product. A method is composed of a series of steps.

There are three points regarding this definition that need clarification. First, a *well defined phase* is one whose inputs, outputs and purpose are well specified, which means that there is no doubt about what the method is to accomplish. Second, although a method is defined as *a series of steps*, this is not intended to imply invariant ordering and does not preclude iterative or even recursive application of the steps. Third, the definition given for method assigns to it a somewhat different and more rigorous meaning than is implied by common usage. For instance, in Software Engineering, where the term is rather loosely and variably applied, it often means the whole approach to system development, as in 'the Yourdon Method' or the 'SSADM Method'. In this book, a method is a way of engineering one phase of development, thereby recognising that different phases may require different approaches and hence different methods.

A *technique* is defined as:

A way of accomplishing a specific task that forms part of a method.

The implication is that the steps in a method are made up of one or more tasks. A technique is therefore used within a method step. A method may rely on many techniques. For example, a step in a method might involve creating a particular view of a system, say an external view of the system's behaviour. One technique that partly meets this requirement is to model the system as a state machine by constructing a state transition diagram (see Chapter 3).

Notations are fundamental to techniques in that they provide the expressive media. However, there is a subtle relationship between a technique and a notation. While the notation will have a unique definition, a technique might leave freedom to use variations of notation that accomplish the same end. For example, a technique 'Real-Time Structured Analysis' may be represented in the notation adopted by Hatley and Pirbhai (1987) or Ward and Mellor (1985). Similarly, formal specifications may be written in Z, VDM or CSP.

Therefore a defined and hierarchical relationship exists between paradigms, methods, techniques and notations, namely: a paradigm operates in phases supported by methods, which use a number of techniques, which in turn are supported by notations.

The position of tools is rather more nebulous, in that they can support the whole, or part, of a paradigm, a method, a single technique, or can act purely as a capture mechanism for a notation. In general, however, tools support methods in that they support and co-ordinate several techniques. For example, the Teamwork CASE tool (Hewlett-Packard 1989) supports Structured Analysis, Structured Design, and Entity-Relationship modelling, etc. Moreover, it supports a number of notational 'flavours' for each technique. Tools should not be taken up until it is clear what they can do and how they should be used, hence they are not discussed further here in a general way. A particular set of tools, which supports the MOOSE paradigm, is assumed throughout this book in the discussions of the methods and techniques that are used in MOOSE. The set of tools is packaged as a single Microsoft Windows application called 'The MOOSE Workbench' (or MOOSEbench).

2.2 SOFTWARE ENGINEERING PARADIGMS

The majority of Software Engineering paradigms reported in the literature have been developed to be broad in application; that is, they can be applied to the development of most types of computer software. Thus, if software is the dominant component of a computer system, these paradigms can be considered applicable to the total system and not just to its software.

The earliest formalisation of a paradigm for software products came in the form of the waterfall lifecycle (Royce 1970), which was expanded upon by Boehm (Boehm 1976). A simplified view of the process is given in Fig. 2.1. The phases, which are followed in sequence, produce deliverables that form input to the subsequent phase. A validation operation is required at the end of each phase to minimise or eliminate the need for iteration between stages, although this does not guarantee that errors will not be discovered in later stages.

This now outdated view of the Software Engineering process has been used extensively for information system development. It establishes some important principles but it has also been widely criticised. One of

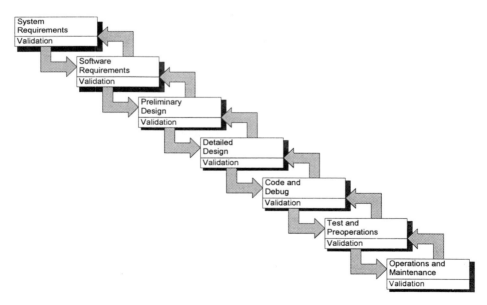

FIG. 2.1 The Waterfall Lifecycle (from Boehm 1976).

the chief arguments against the waterfall paradigm is that there is nothing testable until very late in the lifecycle (Agresti 1987b). In fact an implementation has to be completed before any of it or even the ideas on which it is based can be tested. This can cause a number of problems, the most significant of which is that capturing, analysing and validating the requirements for complex systems is difficult, and errors in the requirements will prove to be very expensive if they are not found until a running version of the software is developed. Specification errors found during the test phase require iteration back through the entire lifecycle, and considerable rework. Boehm's analysis of a number of software developments suggested that the discovery of a specification error in the test phase was twenty times more expensive to repair than if the error had been detected during requirements capture (Boehm 1976).

In effect, the waterfall paradigm places the phases of development in strict time sequence and makes each phase firmly dependent upon the completed results of the previous phase. Nowadays a more acceptable type of paradigm is one that identifies phases of a development process and their purpose, suggests a tentative time ordering for the phases, but allows the possibility of concurrent operation of some phases and iteration between phases. To address these issues a number of alternative paradigms have been proposed. These have the common characteristic of developing a set of executable or analysable models early in the process which, by providing focused and more easily assimilated detail than the typical products of the waterfall paradigm, allow the development team, customers and higher-level managers to review the behaviour of the specification at an early stage.

Two of these proposals involve the development of an executable model of the system as the product of an iteratively applied analysis phase, which is successively refined, with more detail, and reviewed. These approaches are based on Prototyping (Gomma and Scott 1981, Taylor and Standish 1982, Boehm et al. 1984) and Operational Specification (Balzer 1981, Zave 1982, Zave 1984, Agresti 1987a, Agresti 1987b) notations, the significant difference between them being the style of representation used for the model. Prototyping (Fig. 2.2) typically uses a fourth generation language (4GL) as the notation of the model, while Operational Specifications (Fig. 2.3) use a notation geared towards the development of the executable specification. For the former, the model can become the implementation through a process of refinement and the approach is generally used for information systems (Alavi 1984) where the run-time performance of the 4GL implementation is often acceptable. However, a 4GL implementation would not be suitable for reactive system development due to run-time overheads. For Operational Specifications, the model is developed in a specification language and is translated, sometimes manually, into a standard procedural language for execution. The specification notations for such methods are often

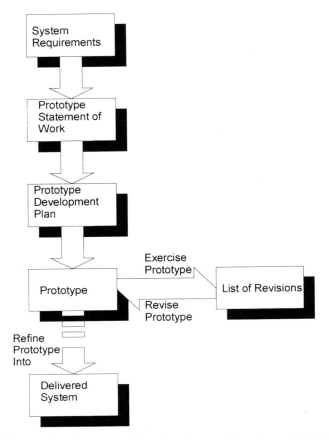

FIG. 2.2 The Prototyping Paradigm (adapted from Agresti 1987a).

designed to address the engineering of reactive systems by providing support for the specification of concurrency and stimulus–response type behaviour; for example, PAISLey (Zave 1982).

A third approach is the Transformational Paradigm (Fig. 2.4) (Balzer 1981, Bauer 1982, Partsch and Steinbrüggen 1983, Agresti 1987c). In this approach a system's behaviour is specified formally, is validated, and an implementation is synthesised by applying a series of transformations to the specification and incrementally adding implementation detail. The advantage of this paradigm is that, if the transformations are consistent, the resulting software is a direct implementation of the specification. However, it has a number of deficiencies, the most significant of which are the problems of satisfactorily specifying real-time behaviour, the difficulties of validating the initial formal specification, the problem of automating the transformation process, and the difficulty in scaling the method to apply it to anything other than individual programs (Agresti 1987c).

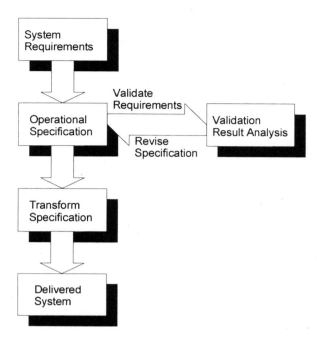

FIG. 2.3 The Operational Specification Paradigm (adapted from Agresti 1987a).

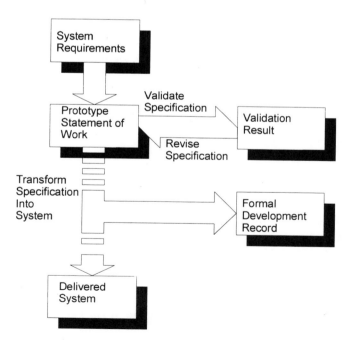

FIG. 2.4 The Transformational Paradigm (adapted from Agresti 1987a).

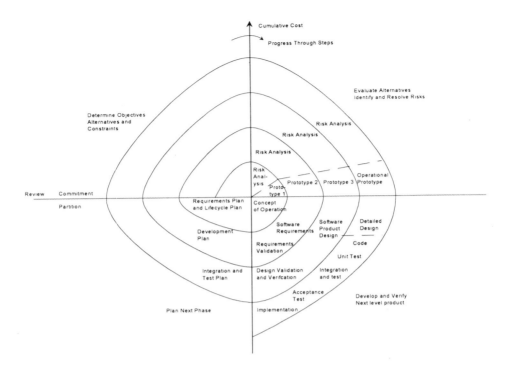

FIG. 2.5 The Spiral Process Model (from Boehm 1988).

The final approach to be considered is Boehm's Spiral Process Model (Fig. 2.5). This paradigm attempts to manage and drive a development project by identifying and controlling major risks at each stage during the development. Rather than fixing upon a particular route for product development in the way that the approaches described above attempt to do, the Spiral Process Model adapts to the risks present in each particular development. This is accomplished by specifying the generic characteristics of a method step, rather than specifying a set of distinct specialised steps. The generic steps are specialised for the particular development project according to perceived risks, particularly in the early stages, and an evolutionary approach results. In particular cases, the Spiral Process Model is equivalent to any of the paradigms described above.

2.2.1 Summary of Software Paradigms

All of the paradigms outlined above attempt, with varying degrees of success, to address three important issues in the development of a software system. These are: that the behaviour of the system can be specified, analysed and validated early in the lifecycle; that critical design issues are evaluated and used to modify the system requirements prior to implementation; and that the transition between analysis, design and implementation is managed in an efficient manner so that consistency is maintained.

The Transformational, Operational Specification and Prototyping paradigms are aimed at the development of specialised classes of software, and while they provide a number of attractive features that can be adapted for a computer system development method, particularly the use of executable models, they cannot be directly applied because they are too software-oriented. The Spiral Process model is adaptable to system development in that it provides the specification of an adaptable method step which can be used in a broad class of applications. However, if this is to be adapted for system development, particular instances of the generic method step and the supporting techniques, notations and tools must be identified.

Many of the software approaches are firmly based on the notion of an analysis phase followed by a design phase, and many sophisticated notations have been devised to express these. Although several aspects of all the software approaches have influenced the paradigm proposed in this book, the activities termed analysis and design are of particular relevance, therefore Chapter 3 is devoted to a discussion of Software Engineering techniques for these.

2.3 APPROACHES TO COMPUTER SYSTEM DEVELOPMENT

Having considered paradigms that are specifically geared to the production of software, we are now concerned with the production of computer systems of which software forms a part. First we must be clear as to the significant difference between a software system and a computer system development paradigm. Stated very simply, for computer system developments, the hardware must be engineered in addition to the system's software. However, we must be careful not to repeat the widely used development paradigm presented in Chapter 1, where the system is divided into software and hardware components at a very early stage, and these are then engineered almost in isolation, each following standard hardware and software development paths. One could consider this approach as 'the engineering of hardware and software components to produce a computer system', whereas more modern approaches to computer system development, including the one presented in this book, are concerned with the engineering of computer systems, some parts of which will be implemented in software and other parts in hardware. Taking this view, the direct application of Software Engineering paradigms and their associated methods to computer system development is deficient in a number of areas.

Firstly, one of the goals of engineering a computer system is to design the *system* in a way that is independent of the ultimate implementation technologies and before determining the hardware–software split for its parts. This means that the methods and notations used must be capable of representing the system and, most significantly, that many of its parts must be implementable in both software and hardware.

Secondly, the paradigm must have a clearly identified step, with associated methods and techniques, that supports the partitioning of the system into its hardware and software components. Furthermore, it must be recognised that there are many forms of hardware implementation for a given functionality that have varying non-functional characteristics, including speed, cost, size, reliability and flexibility, etc. For example, computer system hardware may be implemented as an Application Specific Integrated Circuit (ASIC), in some form of programmable hardware such as a Field Programmable Gate Array (FPGA), a specialised or standard co-processor, or be implemented using standard off-the-shelf components at the macrocell or discrete package level.

Finally, the paradigm must support the identification and design of those system components that do not directly provide the system functionality, but support those components that do. Such components include the system software (the operating system or kernel that manages concurrency in the software and the hardware–software interfaces), the processor and its supporting hardware (including memory and interfacing components such as address decoders and interrupt handlers).

Since 1990 there has been an increasing interest in paradigms and methods for the engineering of computer systems that address the issues presented above. This work has progressed under two titles, Codesign and Computer Systems Engineering (CSE). CSE includes Computer-based Systems Engineering (CBSE) and the Engineering of Computer-based Systems (ECBS). CBSE has been defined by Thomé (1993) as '...the discipline of applying systems engineering to computer based systems', and codesign, by Kumar *et al.* (1993), as '...a concurrent and co-operative design approach that includes, as a fundamental component, the capability to explore hardware/software trade-offs'.

CSE and codesign, from these definitions, are very similar, and indeed many people now use the terms interchangeably. However, historically there has been a significant distinction between their operation and scope. For the purposes of what follows it is convenient to maintain this distinction between the two approaches. CSE practitioners have primarily concentrated on the development of a complete CSE paradigm (equivalent to a Software Engineering paradigm) whose scope encompasses the capture of a system's requirements and specification, through to the system's implementation and testing. Codesigners have primarily focused on one stage of this process, namely the partitioning of a system into its hardware and software components to produce an optimum implementation. In general, codesigners have not been concerned with how the system's specification is developed, and similarly Computer Systems Engineering practitioners have not been concerned with the techniques of partitioning. The following briefly summarises work in each of these areas.

2.3.1 Computer Systems Engineering

We can classify Computer Systems Engineering approaches into two groups, which we shall refer to as homogeneous and heterogeneous paradigms. The homogeneous approaches are concerned with the development of new paradigms, inventing new methods, techniques and notations, although these are generally based on previous work in the Software Engineering field. The homogeneity refers to the notation, in that a single form of representation is used for all models developed. The approach presented in this book, MOOSE, falls into this category. Other approaches that fall into this category are those that seek to use the Statemate tool (discussed later) to engineer computer systems, and that advocated by Woo et al. (Woo et al. 1994).

Heterogeneous paradigms are those that attempt to link together a number of existing software and hardware development methods into a single paradigm. These are of course heterogeneous because a number of notations will be used as the system develops. The term CBSE was developed for methods of this type, and major work in this area has been carried out, for example in the ESPRIT projects ATMOSPHERE and COMPLEMENT, whose respective foci were on systems engineering environments and methods, and on real-time and embedded systems development methods. Results from the ATMOSPHERE projects are published in a number of books, for example Kronolf, Sheehan and Hallmann (1993), Schefstroem and van den Broek (1993), Thomé (1993). In parallel with this work, there have been attempts to establish and promote CBSE through an IEEE task force (Lavi et al. 1991), which is in the process of becoming an IEEE technical committee.

CBSE has a wider remit than simply developing a Computer Systems Engineering paradigm. For instance Thomé (1993) states that:

> Often the term 'systems engineering' is used in conjunction with Computer-Based Systems to denote nothing else other than the engineering of systems containing both software and computer hardware ... However, this understanding of the term is seriously inadequate, as simply a combination of software and hardware engineering, together with some Co-design perhaps, cannot solve the problems which CBSE is setting out to address.

He also notes that CBSE concerns itself with the wider issues of system manufacture, the social consequences of engineering decisions and issues such as the system engineer's responsibility 'to investigate the effects on the natural, economic and social environment of the system'.

While these issues are interesting and must be explored if concurrent engineering techniques are to be developed for computer system products, it is the authors' opinion that the most significant problems concern the engineering of the system from concept, through specification, analysis and design, to implementation and test. Moreover, many of the wider issues are at best poorly understood and at worst a distraction from the very significant problems of engineering computer systems.

The contribution of the CBSE work to the engineering of systems has been twofold. First, generic methods have been proposed (Thomé 1993) that identify the characteristics of a method in terms of its input and output, the activities undertaken and its termination criteria. However, these are not linked into a defined paradigm for Computer Systems Engineering. Secondly, there has been a review of established methods currently used for the engineering of systems and their software and hardware components. Approaches that link these techniques to produce a paradigm for the engineering of complete systems have been investigated. For example, attempts, called 'Method Integration' (Kronolf, Sheehan and Hallman 1993), have been made to integrate methods, such as Structured Analysis with Information Modelling and HOOD; SDL with VHDL; and LOTOS with SDL. Similar work has been undertaken, called 'Tool Integration' (Schefstroem and van den Broek 1993), to link CASE tools. This work takes the pragmatic stance that established techniques and tools that are used in an industrial setting must have some virtue, and that by continuing to use these techniques, companies will not incur retraining costs.

2.3.2 Codesign

Whereas Computer Systems Engineering attempts to try to establish complete paradigms for the development of computer systems, codesign has traditionally assumed a valid specification for the system, focusing on the partitioning of a specification into an optimal hardware–software design and the synthesis of the implementation. The specification of the system can take two forms, either as an implementation independent representation of the system (D'Ambrosio and Hu 1994) or, more commonly, as a representation in an implementation language. In some cases, the latter use a hardware description language such as VHDL as input (Srivastava and Brodersen 1992, Gupta and De Micheli 1993, Kumar *et al.* 1993, Thomas, Adams and Schmit 1993), and migrate part of the functionality from this into software source code, whereas in others a software High-level Language is used as the starting point (Ernst, Henkel and Benner 1993), some parts of which are translated into a hardware description language.

Whether the system specification is in an implementation language or some independent representation, codesign is concerned with providing support, which is generally automated, for the identification of hardware and software components. The goal is to optimise an implementation in terms of factors such as speed and cost, etc., and to satisfy the design constraints. There are a number of ways in which the decision making process is supported; these include performance experiments on specialised prototype systems and performance evaluations to optimise speed (Srivastava and Brodersen 1992, Thomas, Adams and Schmit 1993), and the analysis of system cost factors and the optimisation of these through mathematical techniques such as simulated annealing (Ernst, Henkel and Benner 1993, Kumar *et al.* 1993) or through guided user selection (D'Ambrosio and Hu 1994, Gupta and De Micheli 1993). In most cases the implementation source code of the system is synthesised as part of the process.

An important issue in codesign is the set of implementation architectures that can be used for a particular design. Some methods are deliberately constrained in that the codesign is applied to a particular underlying architecture. These architectures typically have a single, identified processor and also specify the type of application specific hardware, the processor to hardware interface and the operating system kernel (Ernst, Henkel and Benner 1993, Thomas, Adams and Schmit 1993, Gupta and De Micheli 1993). Such methods offer considerable advantages in supporting the partitioning process. Since the supporting hardware and software is known, it is possible to develop accurate cost functions to guide the decision support. Moreover, reconfigurable versions of the architecture can be produced and the designers can test and benchmark their designs running on the target system prior to the final system implementation. Of course, such methods are limited to a class of problems that can be satisfied by the constrained target architecture. To overcome this limitation, some codesign methods offer a much wider choice of implementation architectures by allowing the user to select various processors, styles of hardware implementations and system software (D'Ambrosio and Hu 1994). However, because of the wide range of choices, cost functions and performance models are more difficult to obtain and the process may be less accurate and/or more time consuming.

There is continued interest in codesign techniques of this sort. Results appear promising, particularly for target architecture-constrained problems, but as yet their sophistication is not sufficient for industrial applications (Rozenblit and Buchenrieder 1995).

2.4 SYSTEM DEVELOPMENT TOOLS

In addition to paradigms and methods, there are a number of tools that may be used for Computer System Development. These provide the necessary notations and support for a range of techniques that enable a development paradigm to be defined for the complete system. However, it should be noted that these tools do not enforce a particular paradigm or set of methods, and therefore, while they can be used and have many useful features for the engineering of computer systems, users must identify their own development paradigm and methods of using the tool. There are two well known tools specifically aimed at the development of the complete reactive systems. These are the Statemate tool (Harel *et al.* 1990) and the SES/workbench (Jain, Dhinga and Browne 1989, SES 1992).

The Statemate tool supports analysis, design and the semi-automatic synthesis of implementations. It can be used for system, software or hardware development and provides support for the analysis of functional requirements, timing and, to a limited extent, performance.

Statemate models are captured using three notations that provide dynamic, functional and structural views of the system. System models may be executed to validate functional behaviour, to investigate timing issues and to carry out limited performance studies. Validation of behaviour and timing can be undertaken in a number of ways: through user interaction, where the system developer can view the model's internal and external operation; off-line with files specifying the input data and collecting the system's actions and responses; or through 'brute force' evaluation techniques where each possible combination of input and internal state is tested. This latter feature is particularly useful as it can determine deadlock conditions, verify the reachability of all system states and determine whether there is any unexpected behaviour, although the evaluation may take a substantial time for models of any significant size. Harel provides an example of how this technique was used to determine an unexpected and undesirable sequence of events that could have caused a missile to fire (Harel *et al.* 1990).

Following model validation, design information is carried forward by tools that operate on the model to produce code skeletons for both Ada and VHDL. The tool also automatically develops test stubs for the testing of code.

The Statemate tool appears to be very attractive for the engineering of computer systems and has a number of highly desirable features, particularly executable models and the translation of models into implementation. However, it appears to have a number of problems. Primarily there is no associated development paradigm; guidance on the use of the three notations is rather vague and the accompanying manuals advise the system designer to start in whichever notation is felt to be the most suitable for the particular application. Moreover, there is little guidance on what constitutes a good design, and with the predominant notation being state based, it seems unlikely that optimum structures will be created.

The SES/workbench is less of a computer system development tool than Statemate. Like Statemate, the SES/workbench can be used for many kinds of system at all stages of development, but it is limited almost exclusively to performance modelling. In effect the tool is a graphical programming environment for the creation of C performance models of computer systems, providing extensive run-time support, model animation and sophisticated statistics capture and analysis.

The SES/workbench is an extremely powerful and productive tool for performance modelling, but its proponents suggest that it has a wider role in system development. Indeed, the SES/workbench provides support for translation of the developed model into an implementation notation, VHDL. The tool also provides a useful feature for the validation of hardware designs in a system context, namely the ability of the tool to run models in parallel and to cooperate with hardware models running in a Verilog simulator – so called co-simulation. This allows system performance models to use actual hardware implementations in determining system performance results.

In summary, the SES/workbench is an excellent performance modelling tool that can be used at any stage in the development method to produce accurate results efficiently. However, its use as the primary system modelling tool in a method is unlikely since it is unsuited to both analysis and detailed design.

2.5 MODEL-BASED OBJECT ORIENTED SYSTEMS ENGINEERING (MOOSE)

The MOOSE paradigm aims to provide an 'engineering' approach to the development of computer systems, which achieves a number of important objectives. Pre-eminent is fitness-for-purpose, through good design practice and involving relevant consultation and review. Further goals include ensuring that budget constraints are met, that the time-to-market is reduced (through re-use and synthesis), that efficient use is made of the available skills by prescribing specialised steps in the development process along with procedural guidelines; and the provision of a context for informed decisions with due regard to implementation means and physical realisation. All of this is achieved through a sequence of models. It is recognised that, because the principal computer systems targeted by MOOSE are to be embedded in widely marketed products, competitive pressures will require close to optimum implementation with respect to the performance, response time, physical size, cost and power consumption of the product.

The significance of the term 'model-based' is that the approach operates through the incremental development of a total system model, leading eventually to the automatic synthesis of a high-level language source code for the implementation of both the hardware and software of the system. This is shown in Fig. 2.6. The paradigm starts from a product idea, and the MOOSE-specific part ends when the development can be

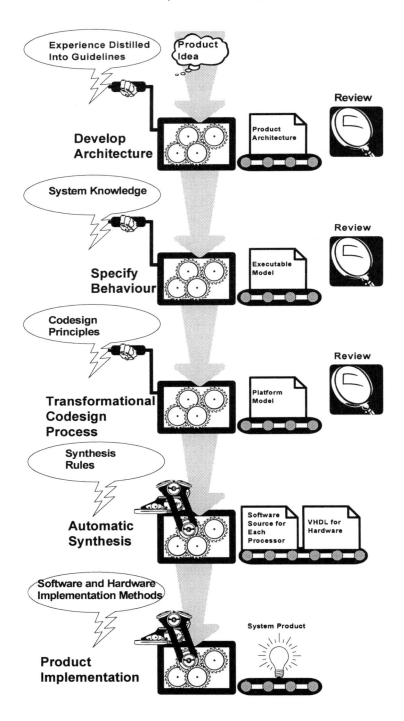

FIG. 2.6 The MOOSE Paradigm.

completed by applying established techniques and standard tools to the high-level implementation source code for the hardware and software parts of the product, these having been synthesised automatically from a 'proven' model.

The structure of the MOOSE models is 'Object Oriented', and this important feature has been found by the authors to provide greater stability of evolving designs, and safer encapsulation of detail, than other structures. However, the notation has to represent objects and the way they interact in an 'uncommitted' style. For instance, objects shown in the architecture may become either hardware or software during codesign, and the style of behavioural definition should not bias this decision. Similarly, the interface of the model should show the logical behaviour but also should not bias the choice of interface hardware nor the design of software dialogue. Thus the features of objects in a MOOSE model are extended beyond those commonly found in software, and a discussion of this, together with a presentation of the details of the MOOSE notation, is given in Chapter 4.

The first three phases defined in Fig. 2.6 are manual operations, albeit assisted by MOOSEbench, concerned with developing an architecture, specifying behaviour so as to produce an Executable Model, and applying Transformational Codesign techniques to the Executable Model to prepare for the synthesis of the implementation. These phases are described in detail in Chapters 5, 6 and 8.

The Platform Model shown in Fig. 2.6 is developed from the total system model to show the relationship between all of the hardware in the system, which includes all the special purpose hardware, processors, buses and interconnect mechanisms, and the software interfaces into this hardware. There is enough information in this model, together with the behavioural detail in the main model, for any style of hardware specification to be synthesised.

For most of this book the implementation language for hardware is assumed to be VHDL and the 'standard tools' will be those that automatically manufacture Application Specific Integrated Circuits (ASICs) from a VHDL specification. Again, this choice is not fundamental but only one of several possibilities, the one that produces the most integrated solution. Other, less sophisticated, technologies for hardware construction were identified in Chapter 1, and interfacing to these technologies, for example through netlists for PCB construction, are considered in Chapter 9.

In the case of software, the object oriented language C++ is assumed to be the implementation language, and the 'standard tools' here will be compilers and debuggers for that language. However, this choice of language is not fundamental to the approach and it can easily adapt to lower-level languages such as C and Assembler, or to alternative object oriented languages. It is also assumed that implementations can use an arbitrary number of processors and the automatic synthesis mechanism produces a separate program source code for each processor, although the objects in different programs will normally interact through a communication mechanism selected by the designers.

2.5.1 The MOOSE Perspective on Analysis and Design

We have already noted that most paradigms have discrete phases early in the process dedicated to analysis and design. MOOSE does not, and so some explanation is appropriate, particularly since the next chapter is devoted to a survey of analysis and design in Software Engineering paradigms. The basic reason for the absence of these discrete phases in the MOOSE paradigm is that the authors believe analysis and design permeate the whole development process and cannot be concentrated in particular phases. To do so is symptomatic of the kind of dogmatic regimentation that underlies the waterfall paradigm, and in any case analysis and design are very difficult to separate.

As Booch remarks, 'the boundaries between analysis and design are fuzzy' (Booch 1994), and other authors have argued an even more extreme view that specification and implementation cannot be fully separated (Swartout and Balzer 1982). Consequently the authors prefer to associate the phases in the MOOSE approach with their deliverables and purpose, rather than with the category of activity that they involve. We have already observed that in any phase of a paradigm we are concerned with a method that can require the application of several techniques. Many of the techniques involve design and analysis effort.

In this book the term *design* is used freely wherever creative effort is to be applied to the development process, albeit in different ways at different stages. For example, it will be seen in Chapter 5 (where the architecture phase is expanded into steps) that, after the Structured Requirements have been developed, an External Model is **designed** to express the externally observable behaviour that is intended of the proposed product. Later, a Behavioural Model is **designed** to show the component breakdown and behaviour, and in parallel a Domain Model is **designed** which places a structure on the domain knowledge. In similar ways, design in this general sense features in all stages of the MOOSE process including those dealing with test and integration policies.

The next question that might be asked is, 'where does analysis fit in?'. Typically the traditional software view of analysis is that it is confined to the initial stages of a development lifecycle. For example, 'In analysis we seek to model the world by discovering the classes and objects that form the vocabulary of the application domain' (Booch 1994). While the philosophy of MOOSE is sympathetic to this view, the authors believe that it is too idealistic to be applied with rigour. In practice, analysis has to interleave with design and both can occur at many stages of the development process. For example, the production of an External Model, whose basic purpose is to model the role of the system in the presence of the application domain objects, requires some analysis of the application domain to be combined with some creative design.

It will also be seen that the activity of creating a Behavioural Model fits well with a traditional view of design: '...in design we invent the abstractions and mechanisms that provide the behaviour that this model requires' (Booch 1994); but, as in most phases of MOOSE, the 'invention' has to be tempered and stimulated by analysis and even by experiments and a feasibility study. Thus we consider techniques of analysis and design to be fundamental to the whole development process and too important to be isolated in specific stages of the development paradigm.

3 Methods of Analysis and Design

During the 1970s the first significant steps were made towards applying 'engineering' techniques, as opposed to 'programming' and 'management' techniques, to the production of large software systems. The 'engineering' techniques that emerged included early versions of approaches of particular relevance to the subject of this book, namely those based on graphical models presenting various views of a proposed system under development. Typically, the views that are produced early in a development are those that support analysis, while the later views identify design structures. Before such techniques came into use, system developments tended to be either programmer led, which meant that nothing existed for external appraisal until working software emerged, or they were controlled from a management level by means of voluminous textual specifications, with poor linkage between the specification and implementation.

By the end of the 1970s, industrial strength 'Structured Methods' had emerged and, although there were many flavours of method from which to choose, certain principles were becoming widely accepted. These included the use of diagrams as the principal form of documentation, and the importance of hierarchical structure and abstraction to achieving clarity. The need to agree what a system should do before deciding how it would do it, the 'divide and conquer' benefits of good modularisation and the value of hiding or encapsulating detail, were also recognised. All were generally acknowledged to be desirable features of a development method. In addition, similarities of approach were evident as the methods usually involved graphical views of a system that depicted the dataflow between the processes, relationships between application domain entities relevant to the system, temporal views in the form of state transition diagrams, and structural views of the software to be implemented. However, the acceptance of principles and the convergence of notational style did not extend so far as to stimulate the standardisation of methods, in spite of some government pressures in that direction.

That so many variants of structured methods exist (see, for example, Demarco 1978, Weinburg 1978, Gane And Sarson 1979) is possibly testimony to the fact the none are ideal, although all are substantially better than not using a method at all. As might be expected, the 'horses for courses' factor was present and the methods best suited to data processing applications were not the best choice for process control applications. Thus the developments of the 1980s led to improvements in methods that specifically targeted particular types of systems; for example, real-time systems (Ward and Mellor 1985, Hatley and Pirbhai 1987), control systems (Bate 1986, Simpson 1986) and information management systems (Jackson 1983, Cameron 1986, Ashworth and Goodlang 1990). These developments were accompanied by the introduction of Computer Aided Software Engineering (CASE) tools on personal workstations, which greatly increased the feasibility of using and maintaining graphical models through the support they provided for error checking, consistency checking, iterative change and even automatic generation of code.

In spite of the attention they received, Structured Methods supported by CASE tools have not proved to be the 'silver bullets' (Brooks 1987)* that some expected them to be, although for a time they were considered the closest approximation available and have been used extensively. As there is a scarcity of comparable techniques for hardware development, the structured methods for analysis and design of software, and particularly real-time software, provide a good base from which to consider corresponding techniques for computer system development.

The fundamental problems with Structured Methods arise from the fact that they are based on top-down refinement. As most authors acknowledge, this is a difficult technique to master, particularly so at the same time as inventing a model of a system that satisfies novel and complex behavioural requirements. An additional problem is that Structured Methods partition the software into modules on the basis of functional decomposition, and there are serious doubts about the stability of these modules when subjected to the changes that inevitably arise as new ideas evolve. For instance, it is not unusual to see a designer disappearing under a sea of discarded diagrams while trying to produce even a first cut of the top-down refinement of a new software system. In fact, some practitioners of Structured Methods (Ward and Mellor 1985, Yourdon 1989) appear to have accepted defeat on this issue and have sought a way around the problem by working middle-out, that is from an intermediate level both upwards and downwards.

Nevertheless, RTSA techniques have been widely used for the development of real-time systems. The Ward–Mellor method in particular provides good pragmatic guidance on the steps to be undertaken in

---

* Folklore suggests that werewolves can only be killed by a silver bullet. In his 1987 paper, Brooks uses the analogy of the problems of engineering software systems as being a monster (the werewolf) and software engineering methods as being the bullets.

order to engineer a software solution for a real-time application, and the combination of notations used produces readable models. However, the Structured Methods have defects which have opened the door to Object Oriented methods. The main criticisms of the former relate to the fact that the initial analysis is based on process (or functional) decomposition and the end result is a design for software that consists of a set of interacting functions that possibly share data. This is a structure that is nowadays considered highly unsatisfactory.

As a result of the observed problems in applying Structured Methods, the proponents of its slightly younger competitor, Object Oriented software development, have found the Software Engineering industry to be surprisingly receptive to their 'new' ideas. Using the sale of CASE tools as a measure, the Object Oriented approach has easily overtaken the Structured approach in popularity in the mid-1990s, for the reasons that will be outlined in Section 3.2. The authors concur with the popular opinion that Object Oriented methods of analysis and design are the best available for most types of software development, and further believe that the main attractions can apply equally to hardware and to systems as a whole. For these reasons, the techniques of Object Orientation must be treated as a source of ideas for a computer system development method, along with those of Structured Methods.

3.1 STRUCTURED METHODS

The Structured Methods chosen for closer examination in this section involve Structured Analysis (SA) followed by Structured Design (SD) (Yourdon 1989), and specialised variants for real-time systems (RTSA) (Ward and Mellor 1985, Hatley and Pirbhai 1987). This choice is based on the widespread use that these methods have achieved for the development of software in systems of the type that we are concerned with here. A number of other structured methods exist that possibly target even more strongly the class of systems of interest, for example MASCOT (Bate 1986, Simpson 1986), and although their use is not so widespread, they are also a useful source of ideas for the computer system developer. We will first give a brief overview of the chosen methods and then present part of a simple model constructed using RTSA.

3.1.1 Structured Analysis

Although there are some differences in style between the variants of SA (and RTSA), the techniques and notations they use have much in common. For instance, all produce a behavioural model of a proposed system through functional decomposition, which expresses the result as a network of concurrent processes (or transformations). In addition they are all centred on the Data Flow Diagram (DFD) notation, which supports analysis and architectural design, to be followed by detailed design based on Structure Charts. In general, DFDs show processes as circles interconnected by lines that represent dataflows. Complex processes are decomposed into simpler processes on subordinate diagrams and the behaviour of the simplest processes (the 'primitives') is specified by text. Shared data is shown by a symbol displaying the name of a data store, usually between two horizontal lines. Read and write access to data from processes is shown explicitly by directed lines. In RTSA, additional symbols using broken lines represent control processes, event flows and control flows. The Ward–Mellor example that follows later should make the principles of this kind of notation clear.

Differences between the methods lie chiefly in the semantic interpretation of some features, the notations and in the specific guidance given to support the system developers using them. In the Ward–Mellor method, the task of developing a design is divided into two major stages: an implementation independent stage called essential modelling, in which the system's requirements are captured and validated; and an implementation specific stage called implementation modelling, in which the system's architecture is developed, leading to Structured Design.

The essential modelling stage starts with an analysis of a system's environment and the required stimulus–response behaviour of the system. A special style of DFD, the context diagram, is used to capture the environment as a single process (in Ward–Mellor terminology, a transformation) connected to externals that provide information inputs and event stimuli, and collect responses. The connections are labelled and their structure and meaning is specified in a Data Dictionary. Stimulus–response behaviour of the system is summarised in a separate event list.

During the construction of the context diagram, a problem often arises with respect to the level of detail at which connections to the environment should be modelled. This type of problem is common to many software and systems engineering techniques, and Ward–Mellor is one of the few to address it. An example of the problem occurs when a connection is to an external system and a multi-layered protocol is required; for example, the OSI 7-layer model. The question arises: what level of the protocol model should the boundary of the system under development (shown on the context diagram) reflect? If the lowest level of software protocol is chosen, the transport layer, the dataflows into and out of the system are concerned with the transfer of packets of data, and so it is difficult to analyse the requirements of the system in terms of the purpose of the communication. If the other extreme of the OSI model, the application layer, is chosen then dataflows that show application-specific features appear on the context diagram. However, as the protocol stack does not appear in the model, its requirements and design could be overlooked during the system's development.

The Ward–Mellor technique addresses the interface problem by assuming initially that pre-processing of information can take place in the external environment of an essential model. Thus, in the case of the OSI communications protocol, the external would supply application specific flows, and the protocol stack for the system under development would be assumed to be external. In the implementation modelling stage, the protocol stack would be brought from the external and placed in the context of the system and the external flows would be shown as being at the packet transfer level.

Following the construction of the context diagram and event list, an essential model is developed as a hierarchical set of DFDs, together with textual specifications of the primitives, the lowest-level transformations. State Transition Diagrams (STDs) are included in the model to define the dynamic behaviour of the control transformations. When necessary, relationships that exist among the data (or information) of the application domain may be documented using Entity Relationship Diagrams (ERDs). However, this data-oriented view is an annotation that is not integrated into the essential model's structure, although guidelines exist for checking consistency between the various views.

In principle, the hierarchy of DFDs is formed by decomposing complex transformations at any level into a group of simpler cooperating transformations, the documentation appearing as a DFD at the next level of hierarchy. This decomposition may be taken to any number of levels and it terminates on transformations that are simple enough to be treated as primitives. However, the process of decomposition is more difficult than might be expected due to the intrinsic problem of identifying the best allocation of roles and function for each transformation, particularly those at the higher levels.

One outward manifestation of this problem is graphically described by Yourdon (Yourdon 1989) as 'analysis paralysis', and it is evaded in both the Yourdon and Ward–Mellor methods by guidelines that propose a 'middle-out' solution. This middle-out approach is based on creating a single (first cut) DFD, containing a transformation for every entry on the event list, then partitioning the transformations on this (rather large) DFD into interconnected groups of transformations that have some cohesive ties. Higher-level diagrams can then be created, which have single transformations that correspond to the selected groups, compatible lower-level diagrams can be drawn to reintroduced the groups but this time on separate diagrams, and the decomposition can be continued on to lower levels in the normal top-down manner. Although it is debatable whether or not the middle-out approach produces a better design, at least it ensures that the modelling process has a defined starting point, and there is empirical evidence that designs started in this way can be completed.

When the SA (or RTSA) model has been developed, it can be reviewed and analysed by walkthroughs (Yourdon 1989b) and, with suitable tools such as Teamwork/ES (Blumofe and Hecht 1988) and Cradle (Structured Software Systems 1994), the model can also be executed according to Ward's execution semantics (Ward 1986). This feature allows the functional requirements to be evaluated fully and some of the timing issues to be addressed.

Development after review continues by 'distorting'* the essential model to produce an implementation model. This distortion takes place in two stages. First, the required processors are identified (these may be microprocessors, human operators or specialist hardware); second, the tasks and their relationships are identified and allocated to the processors. The distortion amounts to a reorganisation of the model, which still remains in the SA (RTSA) notation.

The Yourdon and Hatley–Pirbhai methods are broadly similar to Ward–Mellor. They also have stages dealing with essential and implementation modelling. Yourdon's method (Yourdon 1989) is designed to have broader application than just real-time systems, but his discussion of the modelling of control is limited. Hatley–Pirbhai (Hatley and Pirbhai 1987) provides a modelling notation but little guidance on how it should be used. An interesting feature of Hatley and Pirbhai's technique is the provision of additional notations to express the architectural design of the system, which considers the partitioning of the system into cooperating hardware and software systems. In some respects, the Hatley–Pirbhai technique can be viewed as explicitly supporting some elements of hardware–software codesign.

3.1.2 Structured Design

After an SA (or RTSA) model has been validated and approved, the next step in applying Structured Methods is to translate the SA into a Structured Design (SD) (Yourdon and Constantine 1989). The main component of the SD is a Structure Chart, which defines the software structure as the tree of calls for a set of sequentially executing functions or procedures (sometimes called modules), and it is supported by pseudo-code definitions for each module on the Structure Chart. The task of producing the software may then be delegated, piecemeal if desired, and it may be written in any standard procedural language or, if necessary, in assembly language.

An operational weakness of all Structured Methods is that the validated SA is left behind just when it has reached the peak of its development, and a new structure (the SD) is devised that maps the action of the concurrent processes of SA on to sequentially executing functions. This transformation provides an opening through which errors can enter the

---

* The term is from Ward and Mellor (1985).

design cycle at a comparatively late stage, with potentially costly consequences. However, most methods suggest a set of heuristics and guidelines to aid the designer in producing translations from SA to SD,
although the ease with which this may be accomplished and the quality of
the result are open to question. Good successful designs have undoubted-
ly been produced but it is hard to apportion the credit between design
skill and design method. Another limitation of Structured Methods is that
the focus is on software that will be written in a procedural language, and
those attempts to produce variants that target the object oriented languages (Ward 1989) have not found much favour.

3.1.3 A Structured Methods Example – the Mine Pump

The system chosen to illustrate the application of Structured Methods is
the well known 'Mine Pump' control system, whose purpose is to avoid
flooding in a mine shaft. It has been used by numerous authors to
demonstrate techniques for real-time system design and, although the
behaviour is simple, it has features that bring out the main principles of
a design method.

Figure 3.1 is a schematic representation of the problem. Water collects

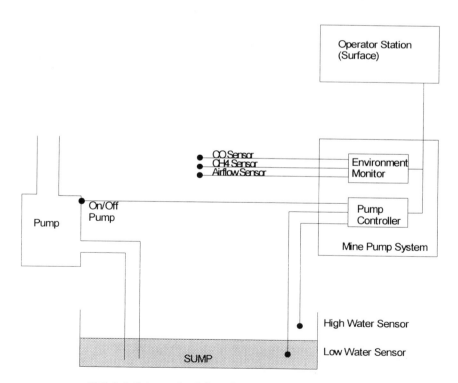

FIG. 3.1 Schematic of the mine pump system.

at the bottom of the mine shaft and when the water level rises above a certain limit (detected by the high water sensor) the pump should be switched on. Similarly, when the water level has been sufficiently reduced (detected by the low water sensor) the pump should be switched off. The pump can also be turned on and off by a human operator. Any operator can control the pump when the water level is between the high and low sensors, and a specific operator, designated the 'supervisor', has the authority to control the pump whatever the water level.

For safety reasons there are sensors monitoring methane (CH_4) and carbon monoxide (CO) concentrations in the atmosphere and the airflow; the operator must be informed of any critical readings, since the area must be immediately evacuated. Further, due to the risk of fire or explosion, the pump must not be operated when the atmosphere contains too much methane. Finally, all three sensors' values along with the pump status (i.e., 'on' or 'off') should be logged periodically for possible future analysis.

3.1.4 RTSA for the Mine Pump

It will be evident to anyone studying Structured Methods that the specification of notations is not always complete and unambiguous, and different versions of their descriptions are not always consistent. This was not a serious problem before CASE tools were introduced, since organisations tended to have their own interpretation and house style. Now CASE tools apply checks and enforce rules, and the rules of a notation are, in effect, those defined by the particular tool. The 'dialect' of RTSA used in this section is the one specific to the SELECT CASE tool (SELECT 1995).

Figure 3.2 is the Context Diagram for the Mine Pump, in which the rectangular boxes represent external devices and the circular 'bubble' or transformation represents the control system itself. The meaning of the diagram should be clear if the reader understands the significance of the three types of lines that are used. The broken lines show event paths, an event being an abstraction of a signal such as an interrupt entering the control system. When an event occurs, an event signal is transmitted and normally some action is expected to follow. Further use of a broken line notation also occurs with control flows and transformations in the lower-level DFDs, as described below. Full lines with single arrows represent paths that convey time-discrete information; i.e., packets of information are sent in the direction of the arrow at a discrete point in time. Full lines with double arrows represent time-continuous information flows, which always have a current value of the information available for the receiver to sample.

Figure 3.3 provides a first-level decomposition of the Pump Control System, hence it is termed the Level 0 diagram. As with all hierarchical refinements of a transformation, the lines that appear to enter the diagram from its perimeter have to be consistent with those that enter and

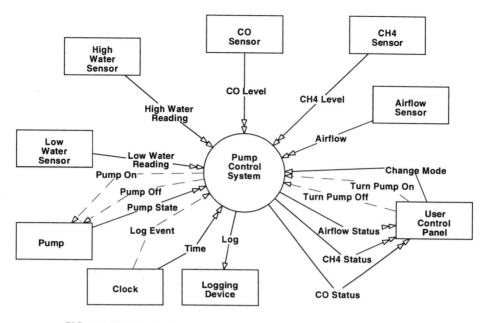

FIG. 3.2 The Ward–Mellor context diagram for the mine pump system.

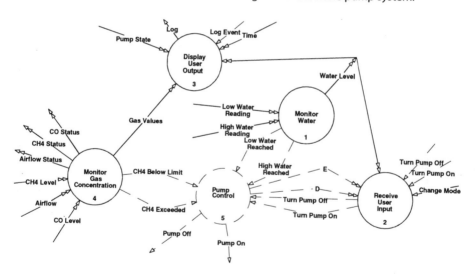

FIG. 3.3 The Ward–Mellor level 0 diagram for the mine pump system.

leave the transformation at the higher level, in this case the Context Diagram. The transformations shown in Fig. 3.3 represent the major functional components of the Pump Control System. The main new aspect of notation used on Fig. 3.3 concerns the control transformation 'Pump Control', shown as a broken circle, whose purpose is to control the

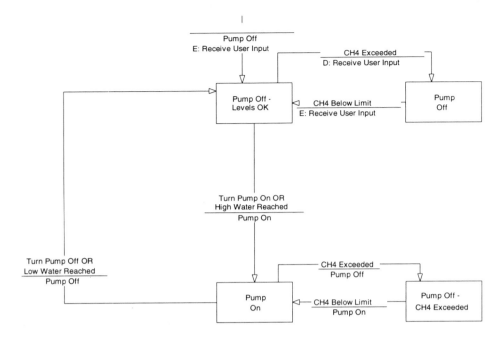

FIG. 3.4 The STD for the control transformation 'Pump Control'.

other transformations. It does this by producing one of three types of control flows. The control flows are broken lines with double arrows and the marking on them (T, E or D) defines their type. A signal on a control flow marked T (a *trigger*) causes the receiving transformation to operate once and then return to an idle state; a signal on a control flow marked E (denoting *enable*) starts a transformation operating continuously; a signal on a control flow marked D (indicating *disable*) causes a continuously operating transformation to become idle. Note that Fig. 3.3 does not make use of any triggers.

Control transformations like the one in Fig. 3.3 are specified as Mealy state machines (Mealy 1955), which receive events and may output event signals and control flow signals as a result of making state transitions. The state behaviour of a control transformation is defined by an STD, for example as in Fig. 3.4.

Readers wishing to examine the complete essential model for the Mine Pump control system should refer to Appendix 2.

3.2 OBJECT ORIENTED SOFTWARE DEVELOPMENT

Arguments supporting Object Oriented (OO) software development fill many publications, but their main thrust is consistent. OO analysis is argued to be the most natural way to understand and model the problem domain in which a system operates, with real-world entities discovered through an analysis of the application domain guiding the choice of objects upon which a system model is based. Moving from analysis to design is also considered to be more straightforward when using OO techniques, since it involves only incremental refinement and development rather than the creation of structurally different models.

In essence, OO approaches produce a refinement of a system in terms of component parts (the *objects*) that have 'crisply' defined interfaces, providing both 'part of' and 'kind of' abstractions (Booch 1994). These abstractions are prompted by the tangible entities and events in the application domain, which eases the problems of discovering them, carrying out the refinement and reasoning about the system.

An important pragmatic advantage of developing an Object Oriented structure is that the interface specification of an object encapsulates its implementation in a way that makes its development and maintenance safer and less error prone. Although the intrinsic difficulty in developing complex systems makes it inevitable that change and iteration take place, Object Oriented structures have a degree of stability not found in process oriented structures.

Particular care is needed in accepting this simplistic view of Object Orientation because, while the benefits of OO methods are substantial, the approach is not as straightforward to apply as some proponents suggest. In many cases the principles behind OO approaches are explained by well chosen systems from nature (for example plants in Booch 1994) rather than by reference to computer software. Although such expositions can make the approach seem extremely attractive and the inferences are seductive, the analogy can obscure a number of pragmatic difficulties. In this book we make no attempt to reproduce the philosophy of Object Orientation and the reader is recommended to study this in some of the referenced texts. Here we will focus on presenting a simplified view of the concept and the salient features of its application.

3.2.1 Object Oriented Programming

The central concept in object oriented programming is the *object*, which may briefly be defined as a kind of software module that presents a crisply defined interface, encapsulating data and associated operations (or methods), relating to an abstraction of an application-relevant entity. Each different kind of object has to be given a *class definition*, which totally defines the behaviour and attribute content of all objects of that class. A program will typically contain many objects, each of which will be an instance of one of the defined classes of object. Each object in a class will have its own copy of state information and attribute values defined for the class, but it will share the same operations as all the other objects in its class. Objects interact by sending messages to each other, which request the execution of one of the operations of the receiving object and supply parameters for the operation. The sender waits until the operation is complete, at which point any results associated with the operation are returned. Objects can be declared statically, in which case they exist for the complete execution time of the program, or they may be created dynamically and deleted by other objects as a program executes.

OO programming has a number of features in common with programming with abstract data types (Liskov *et al.* 1977, Shaw *et al.* 1977), particularly with respect to the introduction of declarative statements that extend the basic type set of a language to include new types whose data can only be manipulated through specified operations. However, the application of object oriented thinking, for example in terms of objects modelling entities in the application domain and the message-passing between objects, leads to a programming paradigm that is totally different from the procedural model.

Lest we conclude that object oriented programming is not as daunting a prospect as it might at first appear, we should note that an object oriented specialist would only regard the programming style implied by the above description as object based, and would argue that the main point of object orientation has been missed. Certainly the description ignores a whole area of significant complexity which is concerned with the mechanisms that can be exploited by the way the classes are defined.

Class definitions can be built up from other class definitions through a mechanism known as inheritance, a term that has been well chosen. If a class definition specifies the use of another class definition, it inherits the definition and may then customise this by adding further attributes and operations of its own. Moreover the inheriting (child) class may redefine operations inherited from the parent class. Multiple inheritance is also possible, in which case the child has several parents and it inherits the properties of all its parents.

Another aspect of inheritance is that the reference to the inherited class may be marked 'private', the biological analogy of which is obscure. In programming terms, the operations of the parent specified by private inheritance are not visible on the external interface of objects that are instances of the child, but the children's own internal operations can make use of the privately inherited operations of the parent.

Some other more complicated features of object oriented programming derive from the inheritance mechanism and concern pointers to any member of a family of objects (in the inheritance sense) and polymorphic behaviour of operations. For the detailed working of inheritance, the reader is referred to an appropriate language description, such as Stroustrup (1991).

3.2.2 Object Oriented Methods

It is more difficult to identify the brand leaders among the Object Oriented methods than is the case for Structured Methods. There is no clear recognition of the 'horses for courses' factor, and many approaches still appear to aspire to becoming the silver bullet. Some interesting but inconclusive debate has taken place on standardisation (OOPSLA 1994) but, meanwhile, signs of convergence in some of the methods can be seen, for example, in the changes that occur in new editions of the documentation for the individual methods. In view of the similarities, we shall make an arbitrary choice of a typical approach from the following shortlist of well known methods, all of which are supported by substantial textbooks and tools available from various sources:

- Object Oriented Development (OOD) (Booch 1991, 1994)
- Object Oriented Analysis and Design (OOA/D) (Coad and Yourdon 1991)
- Object Modelling Techniques (OMT) (Rumbaugh *et al.* 1991)
- Object Oriented Software Engineering (Objectory) (Jacobson *et al.* 1992)
- Object Oriented Analysis (OOA) (Shlaer and Mellor 1992)
- The Fusion Method (Coleman *et al.* 1991)

The 'arbitrary' choice of method we have made is OMT, and this method is used in Section 3.2.4 to demonstrate the principles of object oriented analysis and design, by applying it to the Mine Pump example introduced earlier. However, we first offer some general remarks on Object Oriented analysis and design.

3.2.3 Object Oriented Analysis and Design

The technique of Object Oriented Analysis (OOA) appears to have evolved under three sources of pressure. First, object oriented programming grew out of localised pockets of interest and research specialisation to achieve rapid and widespread usage. The take-up by industry (for example through C++) has produced an urgent need for a specialised 'front end' for object oriented programming, which can serve the purpose that the structured approaches have fulfilled for procedural programming. Second, it is perceived that OO techniques applied to systems analysis, by providing good abstractions of real-world entities, give good *'application domain leverage'* (Coad and Yourdon 1991). Hence an object oriented model is considered to be better suited to the System Analyst's job than Structured Analysis. Finally, the problematic translation between Structured Analysis and Structured Design notations is in principle avoided, since the objects identified and characterised by analysis evolve into the actual objects of the implementation. In addition to removing a source of error (the translation from SA to SD), there is value in the consistency and continuity of technique and notation through analysis and design to implementation. Of course, many other benefits of the OO approach have been observed and are well documented.

In OOA, in contrast to SA, all the methods listed in the previous section use some form of Entity Relationship Diagram (ERD) notation, often heavily adapted, as their chief view of the system. The entity relationship view shows both the relationship between classes and the generalisation–specialisation hierarchy of inheritance. Examples include, in order of conformance to 'standard' ERD notation, the Object Models of Rumbaugh *et al.*, the Information Models of Shlaer and Mellor and the Structure Diagrams of Coad and Yourdon. These entity views are typically augmented by some form of STD to support the modelling of dynamic behaviour, DFDs to show the functional refinement of objects, and some structural views of the system to show the communication between objects and the whole–part hierarchy.

There are two potential difficulties with the kind of notation used by OOA methods; the first concerns problems with ERDs themselves and the second with the number of different notations and views required to model a system. The main problem with ERDs is the size to which models can grow when developing all but trivial systems. For instance, Shlaer–Mellor defines a small domain, that is one that can be analysed as a unit without any attempt to partition the model, as one that contains up to fifty objects. Coad and Yourdon suggest that between fifty and one hundred classes may be held in a sub-domain, although they further divide large class models into subjects to partition the problem domain. Here they suggest the familiar '7 ± 2' rule. The models are essentially flat, which is unsatisfactory because it is difficult to grasp the relationships and associations between classes. For example, if one is to trace through a Generalisation–Specialisation hierarchy, the whole model might have to be scanned. Large models pose many communication problems.

The second problem concerns the number of views and the number of different diagrams that must be created and maintained. A Shlaer–Mellor model, for instance, may have diagrams in as many as seven different notations. Even those techniques that are limited to one notation, such as Coad and Yourdon, do not provide structuring mechanisms that allow the models to be created efficiently, maintained and read without resorting to very large pieces of paper or being restricted to a view of a small subset of the model in a workstation window. In the authors' view, a very major attraction of SA, not evident in most OO methods, is its hierarchical structuring mechanism.

Additional problems concern some of the OO methods themselves, rather than the notation. All provide reasonable support for analysis, but there is no clear route through to design and hence implementation. The guidelines given by methods such as Shlaer–Mellor and Coad and

Yourdon appear superficial. Other methods, such as Objectory proposed by Jacobson, Rumbaugh's OMT and OOD by Booch appear to have better characteristics in this respect and may be considered to be industrial strength methods. However, these methods also suffer from complex and poorly structured models and provide little explicit support for real-time systems.

In summary, although OO methods appear to have considerable merit, computer system developers may still be reluctant to embrace them for a number of pragmatic reasons. These include the fact that many methods, when examined closely, do not offer firm guidance on their application, and so practitioners may be unwilling to jettison their current approaches (perhaps a structured method) which, for all their failings, are at least familiar and well tried. The cost of re-training and re-tooling can also be powerful disincentives to change. Developers may be further discouraged by the emphasis in the literature on the academic aspects of OO.

It might be argued from the perspective of the computer system developer that the most important issue in the early stages of a development is to establish a sound architecture for the system. The classification of components, the relationships between classes and their implementation, are matters to be dealt with in later stages of a development. An aspect of OO that should attract even the most conservative of developers is the promise that it holds out for better re-usability than has yet been realised by any other approach.

3.2.4 Applying Object Oriented Analysis

The application of OOA will now be illustrated using Rumbaugh's OMT to construct a model of the Mine Pump System introduced earlier. Analysis in OMT begins from a problem statement supplied by the clients and possibly the developers of the system. For our example we have the operating scenario summarised in Section 3.1.3 and illustrated by Fig. 3.1.

As with the Structured Methods presented earlier, there are three dimensions to the analysis phase of OMT; these deal with the static structure of the system, its dynamic behaviour in terms of the sequencing of events, and the required functionality with respect to transformation of data. With SA techniques, the latter is of primary importance, as instanced by the pre-eminence of DFDs. These may be linked to STDs, such as by the control processes in RTSA, that specify temporal behaviour, with the modelling of the static structure using ERDs playing a supporting role.

OO techniques take a different stance and usually develop a definitive statement of the kinds of objects required in a system, in the form of a static view of objects and the relationships between them. Using the OMT terminology, this static structure is depicted through an Object Model, with the further supporting views being provided by Dynamic Models and Functional Models. Examples of an Object Model and Dynamic Model can be seen in Figs 3.5 and 3.7 respectively.

A useful starting point in deriving the Object Model is to refer to the textual description of the system to identify the key objects. Thus, a slightly fuller description of the Mine Pump System is given below, in which the names of entities (nouns) perceived to be of relevance to the system are written in bold face (and the verbs are in italics).

Water *seeps* into the **mine shaft** and *collects* in the **sump** at the **bottom** of the **shaft**. When the **water** *reaches* a certain **depth**, the **pump** should be *switched* on and the **sump** drained to allow mining **work** to continue. The depths of **water** at which the **pump** is *switched* on and off are *detected* by a **high water sensor** and a **low water sensor**. The human **operator** may also *switch* the **pump** on or off; a **supervisor** can *switch* the **pump** on or off at any time, a **normal user** can only *switch* the **pump** when the **water level** is not critical.

The scenario is complicated by the fact that **Methane gas** (CH_4) *collects* at the **bottom** of the **mine shaft**, *giving rise to* a risk of **explosion**. Thus the **pump control system** must *monitor* the amount of **Methane gas** via a **sensor** and must *ensure* the **pump** does not *operate* when the **concentration** is above a certain **threshold**. The system also *monitors* the level of **Carbon Monoxide** (CO) and, as it is extremely poisonous, the **mine shaft** must be *evacuated* if a **critically high** level is *detected*.

Critical readings from any of the **sensors** *cause* **alarm messages** to be *sent* to the **operator's console**. The **current status** of the system is *presented* to the **operator** on a **display** and the system also *records* periodically the readings from the **sensors** and the **pump activities** in a **log**.

The technique proposed in OMT for identifying the appropriate classes of objects is based on treating all the entities (nouns) that appear within the product description as candidate classes. In practice, a list based on this technique will include far more than the set of realistic classes for the system, and it needs to be filtered to eliminate redundancy and items that are either vague or irrelevant. Some nouns may also refer to attributes of objects, in which case they are clearly not objects in their own right. Applying this process to the Mine Pump leads to the list given in Table 3.1.

The next stage of analysis concerns the Associations between object classes. These associations are obviously important in showing the interdependencies between classes, such as the conceptual and physical links that exist between objects in the associated classes. For example, the *supervisor* switches on the *pump*. Referring again to the textual description of the system, the required associations can most easily be found by examining the static verbs or verb phrases within the text (as shown in

TABLE 3.1 Classes extracted from problem specification nouns.

Class	Classification
water	Vague
mine-shaft	Irrelevant
sump	Good class
bottom	Irrelevant
depth	Attribute
pump	Good class
work	Irrelevant
high water sensor	Good class
low water sensor	Good class
operator	Good class
supervisor	Attribute
normal user	Attribute
water level	Attribute
methane gas	Good class
pump control system	Vague
sensor	Good class
explosion	Irrelevant
concentration	Attribute
threshold	Attribute
carbon monoxide	Good class
critical readings	Vague
alarm messages	Operation
console	Good class
current status	Vague
display	Good class
pump activities	Vague
log	Good class

italics), which indicate action relationships between entities. This list also needs further refinement to remove redundancy, mainly based on the observation that associations are intended to represent the structural connections between the objects of the application domain, and so consideration of verbs that are clearly steps in a process or that represent a transient event would be deferred until later when the dynamic behaviour of the system is modelled. Applying these principles leads to the list of associations given in Table 3.2.

At this stage, an initial Object Model can be produced which shows the classes prompted by Table 3.1 and the relationships that apply between them prompted by Table 3.2. Further refinement of this model can attach the attributes of the classes to the symbols that represent them to produce the result illustrated in Fig. 3.5.

Initial Object Models go through several stages of refinement, one of which seeks to introduce inheritance wherever two or more classes are observed to have common features, and there is benefit in encapsulating these features in inherited classes. Inheritance possibilities can be discovered in a top-down manner by looking for class names that contain adjectives, such as 'methane sensor'. Since the Mine Pump has several kinds of sensors, it may be appropriate to have a generalised sensor from

TABLE 3.2 Associations extracted from problem specification verbs.

Verb Phrase	Association Classification
Water seeps into shaft	Irrelevant
Water collects in sump	Irrelevant
Water reaches certain depth	Action
Pump switched on	Association between pump and sump
depth detected by sensor	Action
Operator may switch pump	Association between operator and pump
giving rise to explosion	Irrelevant
system must monitor methane	Association between pump and gas
Ensure pump does not operate when	
methane level too high	Association between pump and gas
system must monitor carbon monoxide	Association between pump and gas
shaft must be evacuated	Irrelevant
Critical readings cause alarm messages to be sent	Association between gas and console
status presented to operator	Association between pump and sump and console
system records readings	Action

which specialised sensors such as the methane sensor can inherit properties. An alternative, bottom-up, approach to inheritance is to search
the Object Model for classes that have common or very similar attributes, associations or operations. Neither kind of refinement has been
applied to the Object Model in Fig. 3.5.

The next step in OMT is to develop a Dynamic Model, which shows the
time-dependent behaviour of the system and the objects within it. One
way of approaching this analysis is to identify the events within the sys-

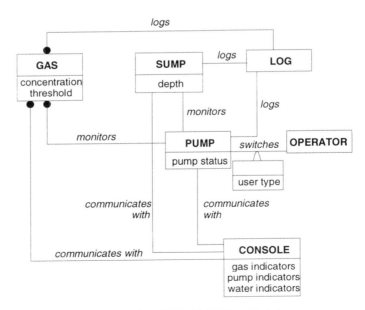

FIG. 3.5 The OMT Initial Object Model.

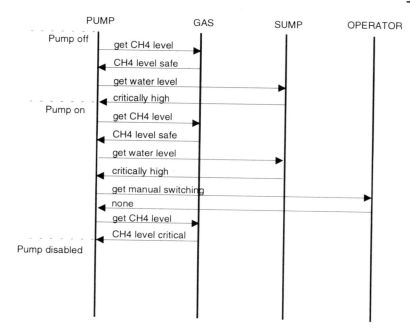

FIG. 3.6 An event trace for an operator interaction.

tem, both between objects and through the external stimuli and responses, and to construct operational scenarios based on typical usage. The logical ordering of events in a scenario can be effectively represented in an Event Trace, as illustrated by Fig. 3.6.

While being a way of inferring logical correctness, there is no mechanism for representing timing requirements, and so it is acknowledged that the method cannot address all the issues of real-time system analysis. From the set of event traces, State Diagrams can be developed for each class. These are only required when a class has significant dynamic behaviour, such as the Pump Class whose State Diagram is shown in Fig. 3.7.

Finally, the analysis phase of OMT addresses the functional behaviour of the system as a whole and a Functional Model is created using DFDs. These diagrams represent the complete system and have a style similar to the RTSA model given earlier. The purpose of the Functional Model is to identify the operations that are required in the classes, and these are then included in the final object model. An example of a DFD for the mine pump is shown in Fig. 3.8.

Of course, the analysis phase is not finished until all three models have been approved, and this may require some iterative change.

As might be expected after analysis, there is a design phase in the OMT method, which incrementally develops the analysis model rather than adopting entirely new notations and model representations, as with the SASD approaches. In principle the transition is intended be seamless,

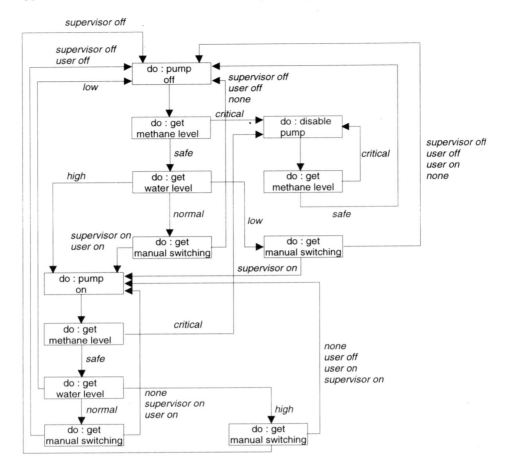

FIG. 3.7 State diagram for the PUMP class.

although there is a difference in perspective as the analysis phases are intended to produce models which are firmly rooted in the application domain and are thus a useful medium for communication with customers and users, while subsequent models focus on computer concepts and the realisation of the system and its components. The design process therefore focuses on issues such as the identification of subsystems and concurrency, and also issues that are important within object design, including data storage and communication mechanisms

3.3 CONCLUDING REMARKS

In spite of the hints of criticism appearing above against OO methods, the weight of opinion and evidence in their favour is overwhelming.

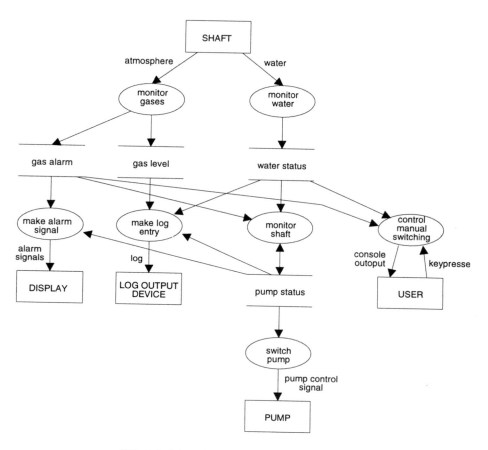

FIG. 3.8 A functional model of the mine pump.

Consequently the notation and method that are presented later in this book seek to apply the principles of object orientation to computer system development. The style of the notation is, however, tailored to fit the mix of hardware and software needs that occur, and to fit a codesign paradigm in which the two aspects of a system are approached in an integrated manner. In its detail, the notation and the method for applying it depart significantly from the object oriented methods described in this chapter, although many of the principles of those methods are adopted. It will also be seen that the notation bears a superficial resemblance to SA, but its semantics and method of use are quite different. However, the hierarchical structure of the models produced owes more to SA than it does to object oriented methods, which are not strong in this respect.

The reasons for developing a new object oriented approach for application to computer systems derive mainly from the 'horses for courses' influences, such as those that have already been mentioned. Our 'course' is the reactive computer systems market, and we need a 'horse' that is equally comfortable in both hard and soft going. Furthermore, the paradigm that we have developed requires that significant design decisions can be made and documented without the need to commit to either a hardware or a software solution for the individual parts. Functional considerations, especially the way in which functional responsibilities are assigned to the component parts of a system, are a critical issue throughout the development process and this is reflected in both the notations and methods used.

In information systems, the *raison d'être* for many objects is to encapsulate the data that represent the attributes of application domain entities and to provide functions that manipulate and control access to these attributes. Suitable structures for many objects can therefore be discovered by entity analysis, and so mainstream OO methods are highly applicable. However, in reactive (process intensive) systems, there is a strong need for objects that play a more active role concerned with implementing system strategies and providing algorithmically complex functions. Entity analysis is not considered a sufficient basis for object selection in such systems, and the authors believe that an alternative method, which focuses more attention on the allocation of functional responsibilities to objects, is needed. Thus a model of the object interactions that take place in a system is central to the method described later. However, it is acknowledged that in all systems there is a need for objects of both kinds, and the paradigm recommends that some aspects of an OMT object model are developed in parallel, specifically to capture the static structure of entities in the application domain.

4 An Object Oriented Notation for Computer System Models

We saw in Chapter 2 that the primary means for documenting and managing the development of a computer system-based product in MOOSE is through models. At the start of the development paradigm, a Behavioural Model.i.Behavioural Model; is created, which defines the required behaviour and structure of the product. After review and approval, this initial model passes through a series of phases of refinement that incrementally accumulate implementation detail. In the first of these phases sufficient detail is added to the model to make it executable, and later phases apply implementation decisions; for example, decisions concerning the use of hardware and software. The executable property of the model is used to construct experiments that support the evaluation of subsequent design decisions. Technical management throughout the product development is based on reviews, positioned at key transition points in the development process, some of which will also exercise the Executable Model. At the end of the development process, high-level language implementations of both the hardware and software for the product are synthesised from the model.

A number of requirements for the modelling notation emerge from the above brief summary. Clearly, the notation must be capable of describing behaviour, in fact it must do this thoroughly enough for the model to be treated as a definition of the functional requirements of a product. In addition, the model has to support the assessments and reviews that form an essential part of a development process. Consequently the notation must have a visual representation that can easily be understood by all those with roles to play in a product's development. For reasons already mentioned, the notation also has to accommodate a specification of behaviour in a computer executable form. The requirement to develop the model incrementally into an implementation means that its modular structure must correspond to the component structure of the implementation. However, in the early stages of a development, components are not committed to either hardware or software, therefore the notation must be suited to both and should not bias the choice that has to be made later. Of course, as the development process proceeds, the notation must allow implementation commitments to be marked on the model, ultimately to an extent that allows implementation source code for both hardware and software to be synthesised from the model.

The purpose of this chapter is to describe the main features of the MOOSE notation (Section 4.3), but the presentation is limited to the features needed for representing Behavioural Models. These features allow system behaviour to be specified in terms of an interconnected set of objects arranged in a hierarchical structure. The form of models used in later stages of the MOOSE paradigm have the same structure, but they employ a small set of additional symbols and markings, which will be described when they are needed. An example of the application of the behavioural modelling notation is given in the form of a MOOSE model for the simple Mine Pump control system that was first used for illustrative purposes in Chapter 3.

This chapter therefore begins with an introduction to the general features of the notation, along with some justification for the form it takes, and this is followed by a firmer definition of the notation. Later sections of the chapter show how the notation is used to develop a Behavioural Model by using the Mine Pump system as an example. In Chapter 5, a more complex model – for a Video Cassette Recorder – is presented as part of a discussion that elaborates the Behavioural Model building techniques.

4.1 FEATURES OF THE NOTATION

For reasons of readability, the MOOSE modelling notation is based on diagrams. However, the software tools that support the development process need to extract sufficient meaning from a captured model to allow them to provide analysis, simulation, synthesis and translation services. Thus the symbols used on the diagrams must have precisely specified significance, and the meaning conveyed by these symbols inevitably needs to be augmented by text. The text linked to the symbols is used both for the informal transfer of information to human readers of the model and for the more precise specifications needed by the supporting tools. For reasons summarised below, the concepts of object orientation in software, suitably extended to cover hardware, form the basis of the graphical notation.

As discussed in Chapter 3, Object Oriented (OO) approaches to software development enjoy strong support in both industry and academia, and OO's main attractions, as the authors expect to demonstrate in later chapters of this book, can apply equally to hardware and to systems as a whole. In essence, the OO approach leads to the partitioning of a system into modules (the *objects*), that have 'crisply' defined interfaces, and which provide both 'part of' and 'kind of' abstractions (Booch 1994). Although abstractions are an important way of obtaining the clarity and understanding that is essential in system design, designing abstractions is usually quite difficult. To some extent, OO eases these design problems because good object choices are prompted by the tangible entities and events in the application domain.

Of course, OO does not remove the fundamental difficulties of developing complex systems, some of which make change and iteration during development inevitable. However, OO structures have been found to exhibit greater stability during such periods of change than most other structures, and this makes the inevitable changes less hazardous. A class definition encapsulates the implementation of its class of objects in a way that hides the detail from other objects, and it forces the users of an object to access its functions and state through a well defined interface. Proper application of OO principles leads to designs with more loosely coupled, cohesive modules than is the case with alternative structuring mechanisms. Finally, class definitions with interfaces providing 'crisply' defined functionality can lead to the build-up of a library of re-useable system components (for both hardware and software) which will ease the task of developing future systems, in the same way that standard integrated circuits (ICs) have eased the task of developing hardware for the past twenty years. The importance of re-use to the achievement of short time-to-market goals was stressed in Chapter 1, but its importance to the profitable development of successful products cannot be over-stressed.

Although the authors believe the attractions of OO are substantial, the underlying principles as they apply to software need some adaptation in order that they can be applied to computer system development. The kinds of roles and responsibilities to be assigned to objects and the ways in which objects can communicate in order to collaborate needs special attention for systems in which some objects are implemented in hardware.

4.2 EXTENDING THE MECHANISMS OF OBJECT ORIENTATION

In software systems, objects collaborate by sending messages that invoke operations (sometimes called methods) in the receiving object, which are implemented as calls to the permitted functions of the receiving object. These messages name the destination object and the required operation, and also supply any necessary parameters. Responses to messages return results. In a graphical representation of a network of collaborating objects, directed lines drawn from sender to receiver remove the need for naming the destination, and annotations on the lines can specify the operation and the parameters (see, for example, Figs 4.1 and 4.2, which are object models for the Mine Pump system introduced in Chapter 3, in a notation attributable to Booch). The notation presented in this chapter will be seen to be similar to that of the Object Models proposed by Booch, but it uses only labels on the lines that name operations, with the parametric details given by associated dictionary entries. More importantly, the notation incorporates a wide range of different line types.

Although it is possible to represent the total behaviour of a mixed hardware/software system by a model in which the only communication between objects is through messages that invoke operations, such a model would have a bias towards the way software components interact, namely through function calls. As a result, an entirely message-based model would not provide a sufficiently uncommitted expression of the communication and interaction taking place in a computer system. This limitation would have an effect at the boundaries of a system, where there may be a requirement both to avoid too early commitment to an actual style and form of interface, and to model precisely pre-determined connections. It also affects the interfaces of those components that are strong candidates for hardware implementation. For example, among the strongest candidates for hardware implementation are those objects that deal with the high bandwidth signals that arise in the analogue parts of vision and audio applications. It would be very inappropriate to model the connections between such components by message protocols. Therefore a purely software style of communication between objects is considered insufficient, both for modelling aspects of behaviour and for providing the necessary flexibility for investigating design trade-offs during the codesign phase.

Another problem with limiting the inter-object communication to messages is that this would not provide an exposure of the critical stimuli that cause the major state transitions within a system, the analysis of which is an important aspect of satisfying time constraints and achieving reliability. For instance, there may be events occurring in a system that are of relevance to several component parts. The object that detects and signals the event may not be aware of which objects have to respond to it. Also, if messages have to be sent explicitly to each object that needs to respond, the signalling object would have the added and undesirable responsibility of knowing to which objects the event is significant, and in what order they are to be informed when there is more than one. The relationship between events and objects that react needs to be explicit at the graphical level, but it should not be present in the implementation of objects. In a similar way, it may be the responsibility of an object to make certain data available on a time-continuous basis without it necessarily being responsible for knowing which objects use the data. Client–server computer systems face a similar situation in that servers do not know the identity of the clients, although the clients need to know from where they can get the service and they have to make a formal request for it.

In reactive systems, the treatment of both events and data typically depends upon system state, which is of course distributed among the objects. This requires that objects can deal with the information flows and events that have relevance for them in their own way and according to their state. Moreover, in some applications, the control of the dynamic routing of information and its transformation, for example its compression, decompression and presentation to a user, is a more important aspect of design than controlling the access to the information. Hence for several reasons the tight control over access provided by message-based structures is too restrictive. In a computer system model, freedom is needed to connect objects to the information flows and events that are evident on the graphical model, without providing operations inside objects to effect the connections.

For reasons such as those outlined, the MOOSE graphical notation for defining object models provides the types of inter-object connections shown in Table 4.1. These connection mechanisms make a distinction between conventional messages passing, continuous data transfer between objects and the propagation of stimuli through an object network.

TABLE 4.1 MOOSE Connections.

Connection Type	Symbol	Time Behaviour	Information Carried	Comms Behaviour	Usage Restrictions
Interaction	name	discrete	data and response data	synchronous	none
Event	name	discrete	none	asynchronous	none
Information Flow – Time Continuous	name	continuous	data	asynchronous	none
Information Flow – Time Discrete	name	discrete	data	asynchronous	To/From External Objects
Parameterised Event	name	discrete	data	asynchronous	To/From External Objects
Bundle of Interactions	NAME	As Interaction	As Interaction	As Interaction	As Interaction
Bundle of Events	NAME	As Events	As Events	As Events	As Events
Bundle of Time-Continuous Information	NAME	As Time-Continuous Information	As Time-Continuous Information	As Time-Continuous Information	As Time-Continuous Information
Bundle of Time-Discrete Information	NAME	As Time-Discrete Information	As Time-Discrete Information	As Time-Discrete Information	As Time-Discrete Information
Bundle of Parameterised Events	NAME	As Parameterised Events	As Parameterised Events	As Parameterised Events	As Parameterised Events
Heterogeneous Bundle	NAME	As Components	As Components	As Components	As Components

Message passing is represented by *interactions* in which one object requests the application of an operation in another object. A receiver is expected to respond to a message, and the response may provide a result from the operation. The mechanism is synchronised in the sense that the sender will wait until the response is received. However, as we shall see later, it is only the function sending the message that waits, and any other concurrent activity in the object containing it will continue.

Information flows, which may be time-continuous or time-discrete, facilitate the representation of data transfers between objects. There is no notion of synchronisation in the case of information flows except that a receiver of a time-discrete flow may explicitly poll to discover if any unread information packets are available. If a producer is running faster than a consumer the packets are queued. A time-continuous information flow might be changed at any time by its producer. Similarly, it always has a 'current' value that the consumers may read.

Events are used to signal the state changes detected by objects that might be of relevance to other objects. The effect of an object signalling an event is that actions may be stimulated in other objects, but the nature and timing of the response is not controlled by the signalling object, nor is there any synchronisation between the object signalling the event and the object servicing it.

A special kind of event, the *parameterised event*, has parameters similar to an interaction that are used when the event is serviced but, unlike an interaction, no commitment is made as to how and when they will be transferred. These events are used to avoid commitment to external interface details in the early stages of modelling system behaviour, as described in Section 4.3.4.

The final type of interconnection is the *bundle*, the purpose of which is to allow a hierarchical structure to be imposed on the connections.

The kinds of communication mechanism introduced above allow a designer to assign to objects the responsibility for: the production or consumption of data on a time-continuous or time-discrete basis; the detection and broadcasting of system relevant events; the provision of responses to events; and, of course, the more familiar provision of operations invoked by messages. Hence they provide abstractions for the common forms of communication found in computer systems.

4.3 DEFINITION OF THE MOOSE NOTATION

A Behavioural Model in the MOOSE notation takes the form of a hierarchy of diagrams called Object Interaction Diagrams (OIDs), in which each diagram represents the partitioning of a 'parent' object from a higher-level diagram into a network of more simple 'child' objects. Appropriate types of interconnections can be introduced to allow the child objects to collaborate in producing the functionality of the parent object. Collectively, the children must have external connections compatible with those of the parent. This section discusses the purpose of the hierarchy, the nature of the objects and the semantics of the connection mechanisms.

The concept of classes and objects was introduced in Chapter 3. We saw that every object is an instance of a class and that the mechanisms for defining classes are extensive and subtle. This chapter is concerned only with objects, in particular the objects that have to be instantiated and connected (or 'made visible') to other objects, so that they may work together to produce the required system behaviour. In effect, the OIDs provide a 'part of' decomposition of the required system and a definition of how the parts are to be connected. Chapter 6 discusses the mechanisms for constructing the class definitions for the objects shown in the OIDs, and some of the ways in which class hierarchies are introduced.

This relegation of classes to an apparently secondary role is just one of several ways in which MOOSE departs from the traditions that are heavily ingrained in OO culture. The primary issues addressed by typical OO approaches are concerned with mechanisms for discovering application-relevant objects, and the means by which the class definitions of these objects may be structured; for example, through inheritance. The justification we offer for this departure is based on the observation that computer systems are mechanisms that are built from components (objects), each of which has a specific role to play in the total system, and collectively as an assembly they must provide the best fit with marketing goals. Thus it is not enough for objects merely to fit nicely into a class; they must also fit the needs of the system. The OIDs, therefore, provide the context in which the latter can best be assessed.

In MOOSE, the view is taken that the instantiated objects in a system, their roles and behaviour, need to be discovered before their classes are defined and appropriate parent classes (inherited classes) are selected. Although, in a more traditional approach, the genealogy of objects expressed through their class structure can have a profound effect on the elegance and re-usability of the class definitions, it is still the instantiated objects that make the product. We are aware that some specialists might argue that the MOOSE approach is object based rather than object oriented, although it is consistent with the 'bottom-up' discovery of inheritance, such as the one recommended in the analysis phase of OMT (Rumbaugh *et al.* 1991).

The emphasis that OO places on class structures in general and inheritance in particular is often justified by the inference that it mimics the structures and classifications that arise in nature, which of course has proven success. However it must be remembered that nature uses natural selection to achieve its optimisation and this technique is not on offer to computer system engineers, as our goal has to be 'right first time'.

4.3.1 The Object Interaction Diagram Hierarchy

As in several other notations for expressing the behaviour of large complex systems, the hierarchical structure is intended to assist in reducing and managing the problems of complexity, by supporting views of a system at different levels of abstraction. At the highest level, a system is represented by a diagram called the **External View**, which depicts the whole system as a single object, surrounded by the external objects with which it has to interface. The connections shown on the External View define the interface between a system and its operating environment. In the example of the Mine Pump system, Fig. 4.1 is the External View. The various styles of connecting lines used on this diagram, in fact all permitted forms of connections on OIDs, are defined in Table 4.1, and the detail of their semantics is described in Section 4.3.4.

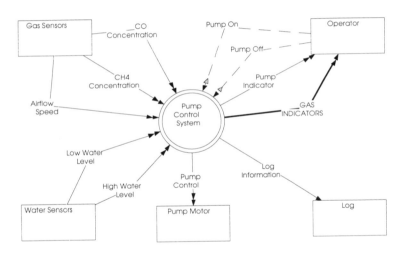

FIG. 4.1 The External View of the mine pump.

The main purpose of an External View is to define the boundary of the system developers' responsibilities and the interfaces the system must have. Thus, all of the system components that are to be specified by the developers of a system are considered to be inside the single system object (double circle) on its External View, and all users and external systems that interface to the system under development are specified as external objects (rectangles). Although an imaginative choice of names for the externals and their connections can imply a great deal about the functions of a system, it does not provide a definition of its externally observable behaviour. This definition of functionality is given by another diagram called the Functional View, which has the general style of a state transition diagram (STD).

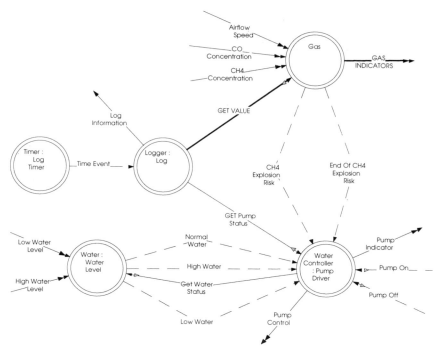

FIG. 4.2 The top level OID of the mine pump.

Criteria to assist in deciding what constitutes 'simple enough' to qualify an object as primitive are discussed in Chapter 5, although there are no hard and fast rules and there is inevitably a strong element of intuitive judgement. Primitive objects clearly do not need child diagrams and, instead, a natural language description of each object's behaviour, called an Object Specification (OSPEC), is linked to all objects that are deemed primitive. At this stage the diagrams, together with their OSPECs, provide the Behavioural Model on which the further development of the system is based. At present we are ignoring the other criterion for terminating the refinement, which is that an object has an interface and behavioural specification compatible with a library object available from a previous development.

Superficially there is a similarity between the appearance of MOOSE models and the Structured Analysis models described in Chapter 3. However, the semantics of the MOOSE notation, which are based on interactions between objects, contrast strongly with those for the passing of data between processes, as in Structured Analysis. In fact the style and semantics of the MOOSE diagrams are clearly much closer to the Booch object diagrams, but with a richer set of connection mechanisms and the ability to accommodate more precise detail so that they can be executed.

In MOOSE, considerable use is also made of a hierarchical structure

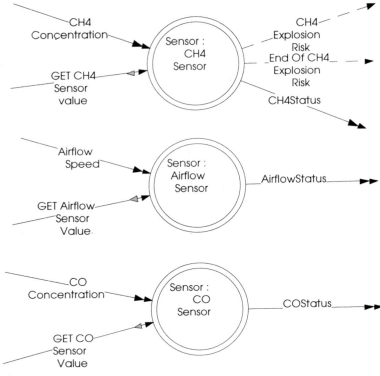

FIG. 4.3 The OID for the **Gas** Object.

placed on the connections between the objects, in a way similar to that applied to the objects themselves. For example, when there are several closely related connections, it is usually better to show them as a 'bundle' of connections on the high-level diagrams, rather than individually. Also, when a bi-directional protocol is necessary to complete a single logical transaction between objects, the readability can be improved by bundling the individual connections and postponing the presentation to lower-level diagrams. Thus the notation for connections allows bundles to be introduced on the high-level diagrams, which give an abstracted view of the role of the object. Bundles are eventually replaced at a lower-level by the constituents of the bundle, which together allow the object to fulfil its role; for instance, the bundle **GAS INDICATORS**, shown in Fig. 4.2, is decomposed into components **CH4 status**, **CO status** and **airflow status** at a lower level, as can be seen in Fig. 4.3.

In summary, the diagram hierarchy allows a Behavioural Model to be structured as a tree. The External View forms the root; the first OID below this, which expands the External View's system object, forms the trunk; the set of OIDs that show the structure of the lower-level system objects forms the innermost branches; the outermost branches are formed by primitive objects; and finally the textual specification of the objects, the OSPECs, forms the leaves.

4.3.2 Functional Views

The hierarchical set of OIDs provides a static view of the product that shows the constituent components of a system and their interrelationships. The Functional View augments the hierarchy of OIDs, providing a view of the dynamic behaviour of the product. As will become clear in Chapter 5, the Functional View is developed in parallel with the External View but, unlike the External View, it has no formal relationship with the OID hierarchy. However, as we shall see in Chapter 5, it is extremely useful in identifying objects and their behaviour.

The Functional View specifies the externally observable behaviour of the complete system by means of a set of State Transition Diagrams (STDs) which, as we shall see later, also have a role to play in specifying the executable behaviour of some objects. Mealy state machines are used, and their basic syntax is described below. However, we should note that, when describing the behaviour of complex systems, the basic notation of STDs must be extended to deal with a number of problems.

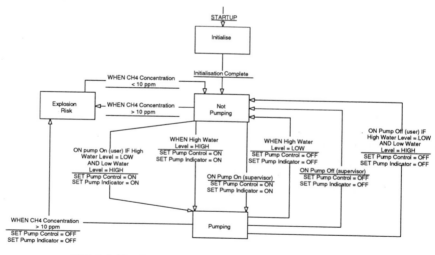

FIG. 4.4 The Functional View of the mine pump.

Figure 4.4 is the Functional View for the Mine Pump System. This diagram shows that the system can be in any one of four mutually exclusive states, represented by the labelled rectangular boxes **Initialise**, **Not Pumping**, **Pumping** and **Explosion Risk**. The system moves from one state to another through the transition arcs, which are shown as directed lines between the states. A transition arc will be taken when an event occurs that matches the condition marked on the arc. The transition condition is expressed by the text above the horizontal line. For example, the system will move from the **Initialise** state to the **Not Pumping** state when the event **Initialisation Complete** occurs. This particular transition is relatively straightforward and it will occur shortly after the system powers up.

Many of the other transitions in Figure 4.4 are governed by more complex conditions. The condition statements are intended to be descriptive and so do not follow a rigorous syntax. However, by convention, keywords are used with the following semantics:

- WHEN. This statement evaluates time-continuous information flow values, generally to compare them with some critical threshold. For example, the transition marked **WHEN CH4 Concentration > 10 ppm** will be taken when the concentration of methane (CH_4) first exceeds the level of 10 parts per million.
- ON. This statement is used to detect events. The transition is taken when the event occurs. If the event is parameterised, the parameters may also be examined. For example, **ON Pump Off (supervisor)** is taken when the user i.d. is that of a supervisor, whereas **ON Pump Off (user)** may be taken when the user i.d. has only standard privileges. Note that, in the latter case, the system also requires additional conditions to be satisfied.
- IF. This is a guard condition. The transition will only be taken if this part of the statement evaluates to 'true'. Thus the transition marked **ON Pump Off (user) IF High Water Level = LOW AND Low Water Level = HIGH** will be taken only when the water level is between the high and low water sensors. If the guard condition evaluates to 'false', the transition is not taken and the event is discarded.
- STARTUP. This transition is taken when the system is entered.

Events that occur when the system is in a state that does not have the corresponding marked transition are ignored and the event is discarded. For example, in the **Initialise** state, incoming events based upon the CH_4 concentration will be ignored.

When a transition is taken, an optional set of output responses can be generated. These responses are marked under the horizontal line of the text annotation associated with the transition. A set of keywords is provided to make the meaning clear.

- SEND. This causes an event to be generated. There is no example in Fig. 4.4.
- SET. This causes a time-continuous value to be changed. For example, **SET Pump Control = ON** sets the named time-continuous value to 'ON'.

When creating the Functional Views of a MOOSE model, it is important to ensure that the names of the states describe the continuous action that the system exhibits in that state. Thus names should be chosen that reflect the state of the system. For example, **Pumping** is a valid state name, but **Turn Pump On** is not; it is properly an event. The names should be chosen so that they are meaningful in the application domain of the system, rather than its design or implementation. For example, **Pumping** should be used in preference to a statement such as **Register 1 = 1**. The same applies to events that cause or are generated by transitions. For instance, **Pump Control = OFF** is preferable to the more cryptic **Pump Control = 0**.

This section has so far described the basic notation for the Functional View. In such a view the system is only ever in a single state and the system is simple enough for all states to fit on one diagram. However, this is insufficient to model the functional behaviour of any but the simplest of systems. A number of features have therefore been introduced to minimise diagram complexity and to represent concurrency. Despite the fact that the Mine Pump system does not need these features, we will introduce them by considering variations of the Mine Pump's Functional View.

The first problem to consider is that of complex and cluttered diagrams; Fig. 4.4 is getting towards the limit of readability on a single page. The solution to this is to allow the sort of hierarchical expansion that was used for OIDs. This introduces the idea of having *State Systems* that are expanded into lower-level STDs. A State System is distinguished from a normal state by having its name given in capitals. Thus, in Fig. 4.5, the states **Not Pumping** and **Pumping** from Fig. 4.4 have been gathered into the State System **NO EXPLOSION RISK**, the constituent states of which are shown in Fig. 4.6.

The second problem concerns concurrent operation within the system. In our discussion of STDs we have limited the system to being in a single state at any given time. However, for many systems this is not practical and thus we have provided a notation that allows a number of State Systems to operate in parallel.

The STDs that have been shown for the Mine Pump capture most of its dynamic behaviour, but they do not address the periodic logging of pump and sensor states that is required in the specification. Once initialisation is complete, the logging operation occurs in parallel with the rest of the system. A log entry is generated whenever the **Log Timer** (Fig. 4.2) object triggers it and this is independent of the state of the rest of the system.

The way in which parallel State Systems are generated is shown in Fig. 4.7, where leaving the **Initialisation** state causes entry to both the **CONTROLLING PUMP** and the **CONTROLLING LOG** State Systems. The State System for **CONTROLLING PUMP** is very similar to that of Fig. 4.4 and is not presented again. The State System for **CONTROLLING LOG** is shown in Fig. 4.8.

As noted above, the Mine Pump system is rather simple, and the example of concurrent operation is greatly simplified by the fact that the

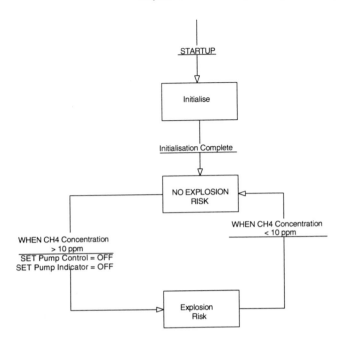

FIG. 4.5 The Functional View of the mine pump with state systems.

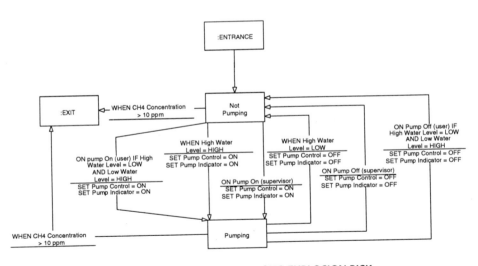

FIG. 4.6 State system of **NO EXPLOSION RISK**.

two State Systems operate in parallel until the system is powered down. In more complex systems, the level of concurrency will change during the operation of the system; at some times there will be a single active STD, at others there will be multiple STDs operating in parallel. Thus

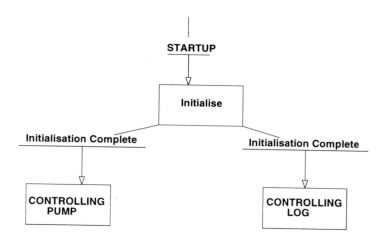

FIG. 4.7 The Functional View of the mine pump with concurrent state systems.

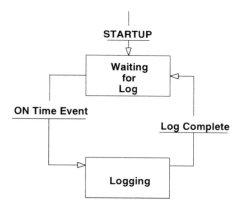

FIG. 4.8 State system of **CONTROLLING LOG**.

there must be provision for concurrent States Systems to be exited. This is done by providing a special state (called **:EXIT**) through which transitions are routed to the higher-level STD. This has been shown in Fig. 4.6.

The adaptations to the notation have been chosen so that the complexity of STDs can be controlled and so that concurrency can be represented. The notation presented here is not the only way in which STDs can be adapted to achieve this purpose. Statecharts (Harel 1987) have the same features and arguably provide more elegant representations. However, the desire to provide readable representations in printed form has led us to use the simple adaptations described.

4.3.3 Objects – Notation

Table 4.2 identifies five kinds of objects. They do not all have distinctive symbols and the differences therefore have to be inferred from other properties, such as the text contained within them.

TABLE 4.2 MOOSE Objects.

Object	Symbol	Restrictions on Usage
External Object	External Object Name	External View ONLY
System Object	Object Name	External View ONLY
Uncommitted Object	Object Name	Any OID (except External View)
Primitive Object	Class Name : Object Name	Any OID (except External View)
Library Object	Class Name : Object name	Any OID (except External View)

External Objects are objects (entities) in the application domain in contact with the system, and they provide the context for the development. These objects are shown on the External View only to provide reference names and to act as the source of interactions, events and data that enter the system, and as the receivers of interactions, responses and data generated by the system. In the Behavioural Model, the behaviour of External Objects is not specified in the same way as that of internal objects, since the external behaviour is assumed to be outside of the responsibilities of the system developers. However, the MOOSEbench tools allow the user to simulate external behaviour by interacting direct-ly with the connections to external objects when a model is executed.

Uncommitted Objects are composites of other objects, i.e. their behav-iour is produced by the behaviour of their 'child' objects, which are pre-sented on lower-level OIDs. The hierarchical structure of the model is built up through these objects. Uncommitted Objects are given a name that is suggestive of their function and responsibilities. The *System Object* shown on the External View is the highest-level Uncommitted Object and, as such, is the starting point of the OID hierarchy. It uses the standard symbol for uncommitted objects and its name is the name of the system; for example, Pump Control System.

Primitive Objects are the lowest-level objects in a system model. They are instantiations of classes for which class definitions will eventually have to be provided, and they have both a class name and an object name. Where a number of Primitive Objects exhibit the same behaviour, they will be instantiations of the same class; hence they will have the same class name but they must have unique object names, as can be seen for the sensors in Fig. 4.3. In addition, each Primitive Object has an OSPEC which, among other things, holds a textual description of its behaviour and the Object Class Mapping. Class definitions and Object Class Mappings are discussed in Chapter 6.

Library Objects are similar to Primitive Objects in that they are not refined into lower-level OIDs within the model, although they may be refined within the library. They are instantiations of a defined class that is taken from the library. There may, of course, be multiple instances of the same library class in a given model. For the purposes of this discus-sion, they may be treated in the same way as primitive objects.

Further details of object notation are postponed until Section 4.3.5 so that the notations for connections can be introduced.

4.3.4 Connections

The types of connection used between objects were introduced earlier and summarised in Table 4.1. There are five distinct types and they may appear singly or be placed in bundles.

Interactions

Conceptually, interactions may be viewed as the way in which one object (a client) uses another object (a server) to accomplish a task in which the server is a specialist. In other words the server provides some operations that are used by its clients. In a number of ways the semantics of an inter-action are similar to those of a software function call, but we must remember that we are not dealing solely with software. First, an interac-tion carries parameters from the client object to the server and, following completion of the operation, a response is returned to the client, which conveys the results of applying the operation if there are any. Second, the effect of an interaction is to synchronise the function that issues the interaction to the response from the server, in the same way that a calling function in software waits until the called function completes. It should be emphasised here that this does not mean that the client object is suspended; it is the particular function within the client that issues the interaction that is suspended.

For some interactions there is benefit in applying prefix conventions to the names; for example, to distinguish data movement – i.e., to GET data or PUT data – from a request to perform an operation. The distinction is useful later in the development when threads of execution and data movements may have to be analysed. Also the distinction might be relevant for its implementation. Of course in a software system there need be no distinction and all three would be implemented by sending a message to invoke an operation. However, in hardware they will normally be implemented through a register interface and the distinction might be significant, for example between passive registers needed for a GET interaction and the control registers that trigger actions.

Time-Continuous Information Flows

The notation for time-continuous information is included so that objects may be assigned the responsibility for making certain information continuously available to any other objects, without being aware of when or where it is accessed by other objects. Examples of time-continuous information flows are found in the Mine Pump model at the interface of the system and its environment. For instance, the concentration of the CH_4 and CO are continuously provided by the appropriate external devices and these concentration levels are continuously monitored by the **Gas** object (Fig. 4.2).

It is important to distinguish clearly between information flows and interactions. Consider a Timer object whose responsibility is to provide 'time-of-day' information to any number of other objects. It could meet this responsibility either by producing a time-continuous information flow or by providing an operation that could be invoked by an interaction. If a time-continuous information flow is produced there is no need for an interaction, and any object may use the time information that is broadcast. Alternatively, an inquiry operation (for example a GETTime interaction) could be used each time the information is required, but the GETTime operation would be called upon repeatedly to deliver the information. The GET interactions constitute an implicit commitment to a software style of information transfer. Time-continuous information flows allow the routing (and possibly the transformation) of such information to be separate from functional interactions.

Time-continuous information flows cater for a wide range of uses and hence actual examples may show significant variation in their bandwidth, rate of change and type. Within the same application, for example, some information flows might be rapidly changing high bandwidth information paths, such as those for carrying multimedia information, while others might be slowly changing, such as time-of-day information. Also, they might be modelling analogue or digital information. The significance of all this variability is that the abstraction provided by the information flows is a powerful one. It separates information flow from functional interactions so that full consideration can be given to the trade-offs involved in committing to specific design decisions. Of course data that represents state information, which is properly thought of as 'belonging' to an object, would not be made accessible in this way, as it is better in this case to control the access by using GET interactions.

Time-Discrete Information Flows

A limitation of time-continuous data is that only its current value is available to the receiver. If it is required that a particular sequence of values of some information is presented to a receiver, Time-Discrete Information Flows are used. Such a flow would be used, for example, when a file was being generated as output from a system. The **Log** object (Fig. 4.2) is connected to the external object **Log** using this type of flow (**Log Information**). It represents the packets of system status information that are periodically sent from the system to an external logging device.

The precise nature of the interface between the logger and the log will be determined in the later Transformational Codesign stage. The Time-Discrete Information Flow is used here to present an abstract view of the communication traffic and requirements. There are queuing implications associated with the packets on time-discrete flows. These can easily be interpreted in a behavioural model but raise issues in implementations with respect to which object should contain the queue. They are therefore only used on the interface of a system and not as a means of inter-object communication.

Event Signals

Events are another necessary feature of reactive computer systems and can arise either in the form of events input by externals or events detected and broadcast by objects. The latter may occur when an object is monitoring a time-continuous data input in order to detect a significant change. This can be seen in the Mine Pump System (Fig. 4.2) where the **Water Level** object produces an event whenever the water level moves beyond a threshold, such as **High Water**, **Low Water** and **Normal Water**, and the **Pump Driver** object responds.

The semantics of an event signal have some similarity to those of an interaction in that operations are invoked in the receiving object in both cases. However, there are significant differences. Unlike operations servicing interactions, functions servicing events may operate concurrently with the functions that issue them. Also, an event signal carries no parameters, the object issuing the event expects no response and the object assigned the responsibility for signalling an event will not know where the responsibility is placed for dealing with it. Thus for events, as with time-continuous information flows, there may not be any natural coupling between sender and receiver. There may be several receivers of an event signal, and in fact several transmitters, hence no restrictions are imposed on the 'forking and joining' that results from event stimuli.

In spite of the differences noted above, the designer of a Behavioural Model may in some situations have difficulty in choosing between events and interactions, and the following guidelines are offered. If parameters or results are involved, or synchronous servicing is required, the option does not arise and interactions must be used. If the intention of an object is to raise an alarm, it should generate an event. Alternatively, if the intention is to request action, it should transmit a request (i.e. an interaction). Finally, if an object has encapsulated state knowledge that undergoes transitions that might be of system-wide interest, it should produce events. Of course the style of name on the line should reflect whether it is signalling an event, broadcasting a change or requesting action, and the impact of this on readability might be used to resolve marginal cases.

Parameterised Events

Parameterised Events are a type of event signals that may only be used in the External View. They are required in situations where an external object, typically a human user, is to give the system a command (or receive an output), but options are to be kept open regarding the interface style. Using an interaction is one way of showing a user command, but this has semantic connotations; i.e., it is very authoritative and synchronous. Sometimes this may be a necessary feature of a user interaction, but in other cases all that may be intended is that a user can signal an event (for example by operating a key) and hope (or at best expect) that the system will notice what is required and take steps to obtain the detail (the parameters) through whatever interface technology is deemed best by the system designer. Choosing such technology would be part of the Transformational Codesign process, as it involves careful consideration of the product's non-functional requirements. The use of a parameterised event instead of an interaction therefore leaves open a wide choice of implementation mechanisms by means of which events can be detected and the associated parameters obtained.

In the Mine Pump model, for example, the **Pump On** and **Pump Off** commands from the user are modelled using parameterised events, which carry the status of the user, that is 'operator' or 'supervisor'. The mechanism for recognising the user status will be developed as part of the Transformational Codesign process and may be, for example, by key lock, voice recognition or some form of password protection. Delaying the decision about user interface style in this way allows the Behavioural Model to act as an architecture for a complete product range, different members of which will have different interfaces, and it therefore ensures that hardware/software commitment issues are removed from the Behavioural Model.

Bundles

Bundles are shown on OIDs as thick lines. If the bundle is made up of a set of connections of the same type, referred to as a homogeneous bundle, it is shown by a line of the appropriate type in bold and with a name in capitals. **GAS INDICATORS** in Fig. 4.2 provides an example of a bundle of time-continuous information flows, and **GET VALUE** a bundle of interactions. If the bundle is made up of connections of different types, called a heterogeneous bundle, the line style used would be as shown in Table 4.1. It must be stressed that the use of bundles to cluster communication detail into higher-level abstractions is just as important as introducing hierarchical abstractions of objects, and sometimes just as difficult to do well.

4.3.5 Objects – Constant Parameters and Object Initialisation

Some objects require configuration information to be encoded explicitly in the model. This is particularly true of a model in which there are several instances of the same class, which have the same behaviour but require different constant values. Consider the Mine Pump's sensors, **CO Sensor**, **Airflow Sensor** and **CH4 Sensor**, shown in Fig. 4.3. Each of these is an instantiation of the same class, **Sensor**, and they operate in the same way, principally comparing an incoming time-continuous information flow with a predefined critical threshold that is constant throughout the operation of the system. However, the different instances have different threshold values. This is achieved by declaring constants in the OSPEC of each of the objects in question. For instance, the **CH4 Sensor** may have a constant THRESHOLD defined as being 10 ppm by writing: THRESHOLD = 10 in its OSPEC. Similarly the **Airflow Sensor** may have a constant THRESHOLD of 1 m/sec and this is indicated by: THRESHOLD = 1.

The name THRESHOLD is associated with a constant in the primitive object's class definition by means of the Object Class Mapping described in Chapter 6.

In this example, each object requires a different value of THRESH-

OLD. However, in some models the value of a constant needs to be common to several primitive objects. Rather than placing multiple definitions in the objects' OSPECs, a value can be assigned to the constant at a higher level in the hierarchy by introducing a 'Definition Node'. These nodes can be placed on any OID and they have a graphical representation that appears as a small pentagon. There is a text associated with these nodes in which any required constants may be defined, after which the constants may be used in any class definition on the diagram on which they appear, or its descendants. If necessary, constants can be redefined at lower levels by either another definition node or in an OSPEC. In general, constants are defined in terms of other constant literals; for example, FAST_SAMPLING_RATE = SAMPLING_RATE * 10 is a permitted form.

4.3.6 Dynamic Object Creation and Implicit Instantiation of Objects

In the description given so far it has been assumed that all of the objects in a system are shown explicitly on the OIDs. If there are primitive objects that are instances of the same class (e.g., the sensors in the Mine Pump example), there is a separate symbol for each with the same class name but unique instance names. This explicit multiple instantiation does not cater for situations in which either objects are created dynamically or there are more instances of a class of object than can reasonably be shown on the OIDs. Clearly the former case is only generally applicable to software and the latter case is more likely to occur in hardware. The MOOSE notation provides for both cases, but the user should beware that their use in a model limits the options available during Transformational Codesign.

Dynamic Object Creation

The notation for specifying dynamic creation of objects requires that one typical instance is shown on an OID for any class of object that is to be created dynamically. Furthermore, the object responsible for creating the dynamic instances must also be shown, and it must be linked to the typical instance by a line type not previously mentioned, a thick dotted line that indicates the parent–child relationship. Implicit in the relationship is the implication that the parent will have access, via a pointer, to any operations of the dynamically created objects, and the pointers may be passed via an interaction to any other 'uncle' object. The parent of an object may also delete it. All of these features derive from the use of C++ constructs in the class definitions (see Chapter 6) of the parents and uncles, but it is useful to indicate in the graphical model where the dynamic creation and pointer linkages are intended to apply. The notation for parent, child and uncle relationships involving dynamic object creation is illustrated by the following simple hypothetical model.

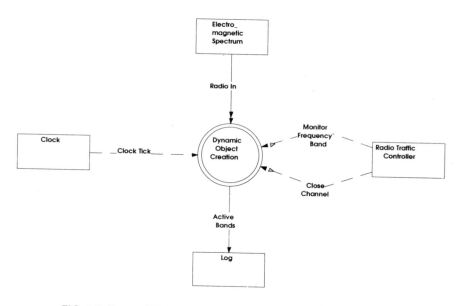

FIG. 4.9 External View of dynamic object creation example model.

Figure 4.9 shows the External View of a Radio Channel Monitoring System, whose purpose is to log the activity of a set of dynamically nominated radio channels. Channels are nominated for monitoring by the external object Radio Traffic Controller through the parameterised event Monitor Frequency Band, and the monitoring of a nominated channel is terminated by Close Channel. In each case the channel is specified by a frequency band parameter. Channels nominated for monitoring have their activity (i.e., the presence or absence of radio traffic) noted in a log entry, which is generated each time the clock event occurs.

Figure 4.10 shows the main components of the system. It uses thick dotted lines indicating a parent–child relationship between the objects **Monitor** and **Channel**, which shows that **Monitor** dynamically creates instances of **Channel**. In fact **Monitor** creates a new instance of **Channel** whenever the Radio Traffic Controller requests it. Similarly, **Monitor** deletes current instances of the **Channel** object when requested to do so. Therefore at any given time there will exist a group of zero or more distinct instances of the **Channel** object, each able to use the **Tuner** object to obtain a measure of the radio activity on the band to which they are assigned.

In this model, the parent object, **Monitor,** is only responsible for creating and deleting the **Channel** objects and it does not interact with the child objects, the **Channel**s, although it could do so if the appropriate interaction line were added to Fig. 4.10. The **Channel** objects are instead used by the **Channel Log** object which, on receiving a Clock Tick,

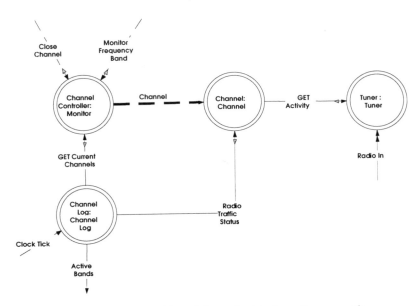

FIG. 4.10 Decomposition of dynamic object creation example.

inquires the status of each monitored channel from the associated **Channel** object. In order that the uncle object **Channel Log** can interact with the dynamically created **Channel** objects, it needs to obtain a pointer to each of them from their parent, **Monitor**. Thus, on receiving each clock tick, **Channel Log** requests the list of current **Channel** objects from **Monitor**, which is able to return the requested list as a list of pointers to the **Channel** objects. Using these pointers, **Channel Log** asks each **Channel** in turn to provide the radio traffic status on its channel.

The Behavioural Model for this example is given in Appendix 5, and it has also been extended to become an executable model, using the techniques described in Chapter 6.

Implicit Instantiation of Objects

In order to provide for very large numbers of statically defined objects, there is a notation for automatic or implicit instantiation of objects, provided that they occur in a suitable repetitive structure. Examples of where the facility might be needed are Artificial Neural Networks, Systolic arrays and arrays of parallel processors. The notation provided requires the whole repeated sequence to be represented graphically by a single symbol on an OID. To indicate that an object is repeated, its object name is followed by a term such as {R1–n}. This implies that n replicas are required and their names will be the name of the group with an integer in the range 1–n appended.

It is also necessary to have conventions for expressing interconnection

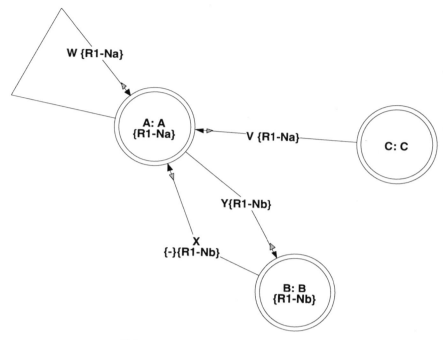

FIG. 4.11 Example of implicit object instantiation.

within and between repeated structures. The hypothetical example in Fig. 4.11 shows two repeated objects A{R1–Na} (of Class **A**), and B{R1–Nb} (of Class **B**), and one non-repeated object **C** (of Class **C**). The repeated interconnections are V{R1–Na}, W{R1–Na}, X{–}{R1–Nb} and Y{R1–Nb}. The interpretations of these interconnections are as follows. Object **C** can invoke the operation **V** of any object in the repeated object **A**. All objects of the class **A** provide an operation **W**, and these objects may invoke the **W** operation of any other object in the repeated class. The interactions **X** and **Y**, which apply between the objects **A** and **B**, have similar semantics. Consistent with the previously mentioned interactions, the **Y** interaction implies that any object in the repeated array **A** can invoke the **Y** operation of any object in the repeated array **B**. However the '{–}' notation in the definition of the **X** interaction provides a mechanism to restrict this interconnectivity, so that only corresponding processes can interact; i.e., object B{i} can only invoke the **X** operation of object A{i}.

The graphical notation indicates the possible interactions between and among repeated objects. However, like any other objects, the actual interactions are specified in the definition of the operations of the classes. Two types of reference to destination objects are allowed, absolute and relative, and the reference can be explicit or computed. This is described further in Chapter 6. The notation generalises to two or more dimensions (Bland and Evans 1994), but we will not give the detail here.

4.3.7 The Model Dictionary

The rules for constructing the Model Dictionary entries are straightforward but, by necessity, they entail the presentation of much detail. In order for the reader to gain an understanding of the nature of the Model Dictionary, some examples from the Mine Pump model are discussed below. The complete Model Dictionary for this model is given in Appendix 2.

The Model Dictionary is used to specify all information relevant to the connections shown in the Behavioural Model, and each connection shown on the model has a dictionary entry. Each entry has two parts: the first part formally describes the connection in terms of its constituents; the second part is an optional comment field that is annotated for human readability with statements that describe the purpose of the connection. The formal part of the entry can be of two types. The first, a *Connection Entry*, presents the structure of an inter-object connection that appears on the model. The second type, an *Information Entry*, defines the structure associated with an Information Name that is used in other entries, as demonstrated below. The most primitive form of Information Entry is a *Descriptive String*, which is a text string that describes the most basic connections or information in the model.

The structure of the Model Dictionary is as follows. Each connection shown in the model has a Connection Entry. If the connection is a bundle, this entry will define its constituents in terms of other bundle names and the names of basic (i.e., non-bundled) lines. The entry for a basic line consists of a set of Information Names or is a Descriptive String. Information Entries may be made up of a set of lower-level Information Names, or they may be Descriptive Strings. Thus the structure is hierarchical and the hierarchical elaboration stops when a Descriptive String is found.

A Descriptive String should not be confused with a comment. The purpose of a Descriptive String is to capture the structural detail necessary to support the definition of behaviour given by OSPECS and, at a later stage, the design of class structures. A Descriptive String is marked by quotation marks ("…") and a comment by the standard C comment markers (/*…*/).

In order to make the structure of the entries clear, a number of examples from the Mine Pump System are given. First, consider the bundle **GAS INDICATORS**. This is made up of the three time-continuous information flows, **CH4 Status**, **CO Status** and **Airflow Status**. This is reflected in the dictionary entry which indicates these constituents and their type:

```
GAS INDICATORS = CH4 Status + CO Status +
    Airflow Status
```

In addition there is an optional comment field that is not used here. In this case the bundle is made up wholly of basic (i.e., primitive) line types. However, in more complex models it is not uncommon for bundles to decompose into more specialised bundles. Each of the entries in the **GAS INDICATORS** bundle is a Connection Entry, one of a group of names that require their own model dictionary entry. Thus there are dictionary entries for **CH4 Status**, **CO Status** and **Airflow Status**. As the information carried by these lines is essentially simple, these are given by a Descriptive String, for example:

```
CH4 Status = "Safe | Danger"
    /* Methane alarm status */
```

This indicates that the status of the methane in the mine can take one of two values, 'danger' or 'safe' (the symbol '|' indicates a logical 'or'). The information in the Descriptive String does not have any formal significance, but is used by those who program the behaviour of the classes that generate and use this information as a prompt to select the appropriate programming language type (see Chapter 6).

In many cases, particularly for interactions, parameterised events and time-discrete data flows, the data carried by some basic lines may be composed of Information Entries that identify the more fundamental components of the information carried. As noted above, these Information Entries have their own Model Dictionary entries. Consider first the entries for the time-discrete information flow, **Log Information**:

```
Log Information = COValue + CH4Value +
    AirFlowValue + Time
```

This is made up of a set of Information Entries, each of which has its own Model Dictionary entry. For example,

```
Time = Date + Hour + Minute + Second
```

is made up of further lower-level Information Entries, such as:

```
Hour = "IntegerRange 0-23"
    /* 24 Clock */
```

This is the most primitive type of entry and there is no need for further expansion.

Our discussion of dictionary entries is concluded by looking at the entries for interactions and parameterised events. These are complicated by the need to specify parameters and, in the case of interactions, to return results. The entries for the parameterised events **Pump Off** and **Pump On** are given below:

```
Pump Off = Operator
Pump On = Operator
```

These simply identify the list of parameters carried by the event; of course, they may also have an optional comment. In the example chosen, both carry the same parameter, which is an Information Entry with its own dictionary entry, given as:

```
Operator = "USER | SUPERVISOR"
    /*Security Class*/
```

The parameters of interactions are identified in the same way as those for parameterised events. In addition, a return value is specified by placing its name between '{}'. The Mine Pump model only has GET interactions that supply no parameters but receive a valid result. For example, the dictionary entry for **GET Water Status** is:

```
GET Water Status = {Water Status}
```

The return value is an Information Entry and may be described by a Descriptive String, which is the case here, or as a further set of Information Entries:

```
Water Status = "High | Normal | Low"
    /* below low water sensor = LOW, between
    high and low water sensors = NORMAL, above
    high water sensor = HIGH */
```

4.4 SUMMARY

This chapter has presented sufficient information for MOOSE Behavioural Models to be read and understood. The next chapter presents a further example and guidelines for constructing such models. Appendices 2–4 provide further example models, including a full Behavioural Model of the Mine Pump System. Later chapters consider how MOOSE models are validated through execution and how they are transformed into an implementation. This will require additional notation that will be introduced at the appropriate points.

5 Developing a Computer System Architecture

Common sense and experience lead us to conclude that an architecture should be developed for a computer system before we set out to build it. However, before discussing the process of producing such an architecture, we need to clarify what constitutes an architecture and how it is to be presented. Like so many other words commonly used by computer specialists, *architecture* is a borrowed word and, for most of us, it has connotations that derive from personal experience, which can easily warp its meaning in our minds.

In the building industry, architecture means the '...design or style of a building...' (Oxford dictionary), but it is doubtful if this definition would make the meaning clear if there was not the widespread understanding that an architecture is produced by an architect, who provides graphic impressions of the intended result to a sponsoring client and working drawings for the builder who is to construct the building. In other words, the term architecture in the building industry has the connotation of documentation produced by those skilled in interpreting the requirements and context of a building, expressing their ideas as a proposal to be approved by a client and a specification to be followed by those skilled in construction. We should also note that an architect will usually rely heavily on models and abstracted views when communicating with the client, and on schematics and other drawings and models that have precise technical meaning when communicating a structural plan to the builder.

This definition of architecture fits our needs very well, provided that there is recognition of the special difficulty of communicating ideas about computer systems, which arise out of their abstract nature, and the fact that this difficulty affects the range and style of models to be provided by the computer system architect. For instance, some models have to convey an understanding of the behaviour that the system will exhibit when viewed from the outside. Other models are needed to specify the structure and organisation of the set of components that will provide the required system behaviour. Both models should stop short of specifying implementation features, since that is a job for the Computer Systems

Engineer who will build the system. However, before the analogy between producing buildings and developing computer systems influences our view too much on where the boundary between architecture and construction lies, we must note some significant differences between the two kinds of product.

With buildings, the role of components will normally be well understood, and so the architecture will specify the positions and dimensions of components and the materials to be used in their construction. Of course, clients will be consulted and builders may be involved; for example, in costing. After agreement on the architecture, the builders will work strictly according to the plan, with little need or scope for making design decisions that have a significant bearing on either physical or functional outcome.

Computer system architectures are inevitably more abstract. Their components are specified in terms of their logical relationships with each other, their functional behaviour and the data they produce, consume, manipulate or transform. There may be some debate as to whether the system architect or the system builder should choose the construction material (hardware or software), but in either case the builder will be expected to apply considerable effort in deciding how each component is to be built. The position, size, cost and performance of components will mainly be determined by the system builder's activities.

5.1 THE MOOSE ARCHITECTURAL MODELS

In the MOOSE paradigm, extensive use is made of models to define an architecture. After these models have been approved, they are transformed by the system designers into implementation models. Thus we conclude that the architect of a computer system should model the functionality that a system is to provide so as to gain user approval, and then produce a model for the system developers to follow which shows how the functionality is distributed among the system components. This implies that the responsibility for determining the best implementation of each component passes from the architect to the developer. This is quite reasonable, however, since the choices to be made depend upon system implementation knowledge and skills. Some decisions may also require extensive analysis and/or prior experimentation with prototype designs. However, the shift of responsibilities raises serious questions about how the developers decide what is 'best', particularly since they are naturally more remote from the user and the application domain than is the architect.

Traditionally it has been the norm in computer system developments for the senior designers (playing the architect role) to decide a hardware structure to which the software must fit. The architecture consists of diagrams that show the functional components and the way they are to be interconnected. The building of the hardware and software then proceeds independently, the latter being done to a requirement specification rather than an architecture. In both cases, standard components will be used wherever possible, but many will have to be engineered specially.

Large scale information system developments typically follow a different development path, in which an abstract model of the required information processing system is constructed, such as a Structured Analysis (SA) model (Yourdon 1989) or a Logical System Specification in SSADM (Ashworth and Goodlang 1990). From this model, the software design is developed and the hardware requirement is determined. Normally for such systems, the policy is to buy general purpose hardware, rather than to build custom hardware, and the only application-specific components are in the software.

Similar approaches are used by the Object Oriented paradigms – for example OMT (Rumbaugh *et al.* 1991) – although these may start the development at a higher level of abstraction, with models of the application domain objects relevant to the system and an identification of the system functionality to be built into their class definitions. However, in both SA and OMT, the underlying principle is to identify what the software has to do and how it is to be structured before embarking on its implementation. The models produced thereby provide an architecture for the software that is consistent with our definition, but they are not required to contribute to the hardware development thread.

Returning to the computer systems that are the focus of this chapter, their complexity has now reached a level at which the traditional approach is unsatisfactory. In most cases, software plays the dominant role with respect to controlling the functionality of the system, which is where the complexity peaks, and a hardware-led architecture would be inappropriate. However, the hardware may be required to implement complex functionality, particularly where it is time-critical, and the treatment of architecture must allow for this. Thus an integrated approach to hardware/software development is needed from the architecture level downwards, with their respective developments being closely linked in a manner that makes the division between them seamless.

Decisions must be taken at the appropriate time and stage in the development process. They must make use of the appropriate specialised skills, and they must be properly informed. For example, at the time that the decisions on partitioning the functional responsibilities between hardware and software are made, they must be based on agreed criteria that can lead to optimum cost/performance trade-offs and compliance with other non-functional requirements, such as size and power consumption constraints.

The architecture phase in MOOSE is the first major step in the integrated development paradigm. It delivers a model for the complete system that encompasses the total functionality that is to be provided, without making specific commitments to the roles of hardware and software, and it includes the results of requirements analysis and application domain analysis.

Chapter 4 introduced the MOOSE notation for creating Behavioural Models of computer systems, in which the overall functionality of a system and its breakdown into functional components are fully specified. A model constructed in this notation fulfils part of the requirements of an architecture, in that it specifies the external behaviour of a system and provides an internal decomposition into implementable components. However, since we have argued that the architecture should provide all the information necessary for the system builders to produce a successful product, two other items are required in the architecture.

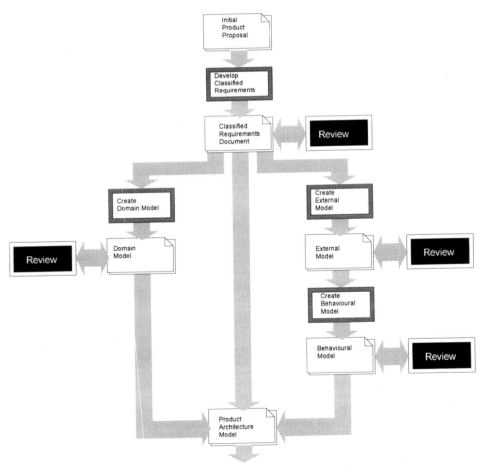

FIG. 5.1 The architecture phase of MOOSE.

First, while the Behavioural Model adequately covers the functional features of external behaviour, and these determine the user appeal of a product in a way that leaves flexibility of choice with respect to the physical interface mechanisms, the buyer appeal of the product – and hence its commercial success – will depend strongly on the extent to which the design optimises non-functional characteristics. For example, the price, size and possibly the power consumption (battery life) will be of significant interest to a discerning buyer. Also, any product with inadequate performance or speed of response will sooner or later lead to serious customer dissatisfaction. The results of an investigation of the non-functional requirements need, therefore, to be included in the specification of system architecture that is passed to the system's developers, to provide criteria that will allow optimum design and implementation choices to be made.

Second, all computer system developments need to take account of application domain knowledge in the course of transforming a Behavioural Model towards an implementation, albeit to a varying extent that depends upon the characteristics of the system. For example, some systems may need to store and reference significant amounts of information relating to application domain entities. There may also be a changing population of application domain entities, modelled using dynamically created objects, which other objects in a system are required to manipulate. In order to carry forward the appropriate application domain knowledge from the system architects to the system developers, an architectural specification needs to include a Domain Model. A number of OO software engineering paradigms include this concept, including OMT which will be illustrated later.

Figure 5.1 summarises the process for producing the three parts of the MOOSE architecture, namely its Structured Requirements, Behavioural Model and Domain Model. The remainder of this chapter discusses the process, and illustrates it with reference to a hypothetical project to develop a Video Cassette Recorder (VCR) product.

5.2 ANALYSING AND CLASSIFYING REQUIREMENTS

The first step of the Architecture Phase of the MOOSE paradigm concerns the analysis of an Initial Product Proposal. In practice, such proposals arise in many ways; they may be the result of a perceived market opportunity, an invitation to tender, or the recognition of a need for a specialised subsystem to enhance an existing product. With this variability, it is likely that the product proposal is deficient in detail. The problem is not unique to computer system-based products and, in an information system development, it would be addressed by a requirements capture phase in which all knowledgeable views of the system requirements and user views in particular would be sought.

However, some of the most attractive and worthwhile computer system developments are those that anticipate and create new markets before they have been widely recognised. In such cases, informed views and experienced users will not exist and some creative design effort will be needed before the proposed product can be specified fully. In the MOOSE paradigm, this creative design effort is applied through models and it is not expected that it will be captured in the more traditional textual requirements specifications. Thus the development of a requirements document, as shown in Fig. 5.1, produces a textual result that is largely confined to classifying and clarifying what is initially known about a proposed product, with only marginal additions to approach completeness.

The classification of requirements distinguishes the Functional Requirements, which provide the primary goals for the designers of the Behavioural Model, from an assortment of other requirements that we shall call Design Constraints, because their effect is to limit the design space in which an implementation can be developed. At this stage in the development, the functional requirements will usually only be given as an interim summary and the main review of functionality will be made against the Behavioural Model, with some functional issues even remaining undecided until validation experiments and feasibility studies can be conducted with an executable model. Within the Design Constraints we identify the subcategories of Non-Functional Requirements, Design Objectives and Design Decisions.

A classification of this kind was first proposed for the requirements analysis of software systems (Easteal and Davies 1989), and it is a particularly useful classification for MOOSE since the categories it introduces have specific roles to play at different stages of the paradigm.

The Non-Functional Requirements specify a range of targets for performance, response time, reliability, cost and size. These become very significant when the means for implementing a system are considered. For example, how else could a designer decide between implementing a particular function in hardware or software? Should a result be computed rather than looked up in a table? Of course, this assignment of responsibility raises another question, namely, how is a designer to establish that a particular choice or set of choices meets the Non-Functional Requirements of the system? Chapter 7 examines questions of this kind in some depth, and later chapters discuss the ways in which the MOOSE tools can assist the designer who faces them.

The Design Objectives usually relate to test, maintenance and general performance requirements that cannot be quantified precisely. They are normally expressed in a style such as 'must be as fast as possible', 'must be inexpensive' or 'must be easy to extend'. In some cases it may be possible to convert such statements into Non-Functional Requirements by agreeing, for example, a price goal or a performance target, but otherwise they should be taken as the selection criteria for making choices between functionally equivalent alternatives when there are no other firm criteria.

The Design Decisions contained in a statement of requirements for a software system are usually considered undesirable, and a common recommendation is to negotiate their removal.

> ...it is the type of statement that is often encountered if the user is relatively experienced in the use of computing techniques ... the user is taking a design decision prematurely. In a nutshell, the wrong person is taking the wrong decision at the wrong time and, for all we know at this stage, sowing the seeds for future disaster. (Easteal and Davies 1989)

In computer system development, while Design Decisions in the requirements should still be questioned, they will often be found to be inescapable. This is because such a system may be a member of a family, a part of a larger product, or it may need to conform to company policy or house style, and these pragmatic considerations are bound to impact upon design freedom. Some design decisions may influence the functionality to be provided, while others only affect implementation choices. Therefore Design Decisions may have relevance at several points in a development paradigm.

When a classified list of requirements has been prepared, they should be agreed with the sponsors of the project, after which the Functional Requirements component is developed into a Behavioural Model. This model must be consistent with the Design Constraints, and Design Decisions in particular might affect the external behaviour of the model. For example, if a particular bus is specified for an external connection, it will normally appear in the External View diagram.

5.2.1 Classified Requirements for the VCR Example

For our hypothetical product development project, we shall consider the proposal for a new VCR product given below. The MOOSE policy for producing Structured Requirements assumes that such proposals are read with the expectation that they may contain errors and inconsistencies, and judgements to remedy these deficiencies are to be applied in producing a classified list of requirements.

> Marketing surveys have indicated that it will be profitable to launch a new range of low cost, easy to use VCRs, aimed primarily at those who find that most typical VCRs offer too much functionality and have complex interfaces. A second market is those who wish to have a cheap, second VCR to use with a portable television outside a home's main viewing room; for example, in a child's bedroom.

Market research suggests that the product should have the normal basic features of a VCR, in that it will provide playing and recording facilities, together with a simple tape transport (i.e., rewind, fast forward, cue and review). The system should also have a seven-day clock, which displays time and day at all times, but date is not required. Limited facilities should be provided for pre-set recordings, which allow up to two separate programmes to be recorded up to a week ahead by specifying the day, the hour, the minute, the length of the programme and the channel number required. More sophisticated facilities often found in other products, such as recording at the same time every day for a specified number of days, are to be avoided in the interests of promoting a 'simple to use' image.

Price and packaging are thought to be the two other most important factors to the product's success. Thus system production and component cost must be minimised. To this end, an existing mass-produced tape mechanism will be bought in. Size and weight are also important issues as the product may be used as a 'second machine', and so a small, portable system is required.

The VCR system is to be connected between a standard television set and its aerial. It should also have two modes of operation; active and standby. In the active mode, all the functions of the VCR should be active. In standby mode, only the clock should be operating, the aerial should be routed directly to the television and power should be removed from all other functional units to reduce consumption to a minimum.

From the above, and applying some judgement as suggested, the following classified requirements might be compiled.

Functional Requirements

The required VCR must connect between the Aerial and Television on the UHF cable. While powered up it may be in one of two principal modes; 'active' or 'standby'. In standby mode, only the clock will be active and the UHF signal will be routed directly to the Television. In active mode the machine will play tapes and provide operations concerned with getting to the selected points on a tape. A seven-day clock will give time and day, and facilities will be provided for setting it. Similar facilities will be provided to allow up to two times to be specified for automatic recording to be made on to tape, from specified channels and for specified periods of time. An 'auto record mode' should be selectable (and cancellable), in which the VCR is not available for any other use. A facility must exist to set the tuning of pre-set video channels; i.e., specification of the frequency to be associated with a video channel number.

Design Constraints

There are three *Design Decisions* in the VCR proposal, one of which is quite clear in stipulating the use a standard Tape Drive. Perhaps less obvious is the decision limiting the number of pre-set recording requests to two, and defining their content as day/time/duration/channel. Finally the explicit restriction against adding functionality is arguably a third Design Decision, albeit of a rather negative kind.

The *Design Objectives* are to produce a robustly packaged product that is small in size and weight, in order to make it easy and convenient to move from place to place.

Finally, there were no *Non-Functional Requirements*. The low cost objective was considered too vague to apply, and a limit of £40 has subsequently been agreed on the manufacturing cost.

Treatment of Classified Requirements

As can be seen in Fig. 5.1, the classified requirements are subject to the approval of the sponsors of the product, which may result in changes before they are regarded as suitable to steer the subsequent steps in the product development. After this initial review, the Functional Requirements are developed in the form of the Behavioural Model. They have no further relevance, although the functionality to be provided by the system may be changed as a result of reviews based on the Behavioural Model.

Especially careful consideration needs to be given to the Design Constraints since, in general, they will influence whether or not a product will be fit for its intended purpose and market. Consequently they can have relevance to all phases of a product's development. The Design Constraints are therefore passed on from stage to stage as the development proceeds.

In the case of the VCR, the Design Decision concerning the tape drive has immediate effect and the specified tape drive is introduced as an external on the External View. The decision regarding details of the pre-set recording is not used until an Executable Model is produced.

5.3 CREATING A MOOSE BEHAVIOURAL MODEL

The diagrams used to specify a MOOSE Behavioural Model are intended to bring out clearly what a system is doing without imposing constraints on how it is to be done. We have seen with the example of the Mine Pump model in Chapter 4 that these diagrams show the component objects as 'bubbles', and communication between them as directed lines of various kinds. As with most other graphical methods for designing large complex systems, a major part of rendering the design problems tractable is through the development of abstractions by way of hierarchical structures. Thus 'large conceptual chunks' of the system appear as objects at the higher levels of the diagram hierarchy, and these are progressively refined into simpler objects, on subordinate ('child') diagrams. When objects are reached that are simple enough to be treated as primitives, their contribution to the function of the system is given by a textual OSPEC.

As can be seen in Fig. 5.1, before a Behavioural Model is developed, an External Model consisting of External and Functional Views is produced and reviewed. The purpose of this model is to specify how the computer system will appear to its users, thereby clarifying the terms of reference for the designers of the Behavioural Model. To ensure consistency, the Behavioural Model uses the External View from the External Model as the top level of its OID hierarchy. In principle, the production of the remainder of the Behavioural Model is straightforward. However, the process is based on top-down development, which is never easy, and a defined approach (in the form of guidelines) is needed to ensure good results.

Before elaborating further on these guidelines, we must emphasise that they will only supplement experience and natural flair and are no substitute for these attributes. However, those experienced in computer system design will be aware of the extent to which these natural skills are stretched by the vast amount of complexity that exists in the design of even a moderately sophisticated computer system. This is particularly true with respect to a system's behaviour in the time dimension. If we do not have guidelines, then we are forced to accept that there is some mysterious art that guides good designers. Judging by industry statistics on its success rate in computer system development, this art can only be known to a small number of practitioners.

In the MOOSE paradigm, the art of design would need to sustain the invention of a network of communicating objects, in which each object represents a good (application relevant) *abstraction* and provides *encapsulated* functionality that has strong *cohesion* and low dependency on external design detail (or *coupling*). It is no more realistic to rely on finding artists with the required talents than it is to expect success without having appropriate skill and experience available. Guidelines are therefore needed to harness our limited skills and accumulated experience. Such guidelines for creating a Behavioural Model are explained below in the context of the hypothetical product development for the VCR, but first a warning of a trap for the unwary!

Throughout this book, the methods are explained with the aid of simple and possibly familiar examples. Their complexity is slight by comparison with realistic systems and the impression might be formed that the guidelines are excessive and superfluous. However, those who either teach computer system development methods or rely upon them to control product developments will readily appreciate the problems of not having guidelines; the case histories of some complex computer system developments presented in Chapter 1 also support this view.

Finally, to make a general point about guidelines, we would like to stress that, while we are in no doubt about the need for the guidelines, it is more important that they are *strong* guidelines rather than *good* guidelines. If the inconsistency between models produced by different designers is eliminated, bad designs can be traced back to guidelines that need improvement. We have at present no evidence that the guidelines are the optimum, but they have been used with significant success.

5.3.1 Guidelines for the External Model

As stated, the External Model is the first step in producing a Behavioural Model. Input to the task is provided in the form of the Classified Requirements and its output comprises the External View and Functional View diagrams, supported by a textual definition of the detail. With the review by sponsors in mind, the terminology used in the External Model should be that of the application domain so that the reviewers need not be computer specialists. Satisfying the Functional Requirements is the most significant part of the task, but the Design Constraints must be strictly observed.

The External View is an OID, which depicts the complete system under development as a single object surrounded by the external objects with which the system needs to communicate. After gaining approval as a satisfactory definition of the external interface, this diagram becomes an integral part of the Behavioural Model.

The Functional View presents a state machine view of the functionality to be experienced by a user, and its purpose is to document the required functionality. The Behavioural Model produced later has to exhibit the same functionality, but there is no formal link with the Functional View.

Assuming that the errors, ambiguities and omissions in the Initial Product Proposal have been addressed while forming and reviewing the Classified Requirements, it might be assumed that it would be a straightforward task to model the specified external behaviour in a graphic form. However, achieving an appropriate level of abstraction and a suitable structure usually proves difficult, particularly if it is recognised that abstraction should not be confused with vagueness and approximation. Detail can only be brought into the model through refinement of what is implicit in the high-level abstractions, therefore, while being abstract, the External Model has to be comprehensive. The essence of the guidelines for producing the External Model is that the development should proceed through the following steps.

- Construct lists of relevant entities, stimuli and responses.
- Produce the External View.
- Specify connection detail in the Model Dictionary.
- Produce the Functional View.
- Review and refine as necessary.

Guidelines for Entity, Stimulus and Response Lists

This first action towards producing an External Model should be based on visualising the system operating in its application domain. A list of all the entities in the application domain that are of relevance to the system should be compiled, and from this list – and with the application of some imagination – the stimuli that the entities might produce and their expected responses can also be listed. This activity overlaps the Domain Analysis that is shown as an independent activity in Fig. 5.1, but clearly some interaction and transfer of ideas between the two will be advantageous. Separate lists for the three items are all that is required, and the temptation to arbitrarily associate entities, stimuli and responses is best avoided, since these associations can quickly become complex and confusing.

It is useful with respect to the stimuli, and possibly also the responses, to identify an ordering. There will usually be 'context' groupings of stimuli; i.e., groups that only have relevance while the system is in a specific state. Also, it will usually be possible to recognise the stimuli responsible for entering these states, and in some cases 'complementary' stimuli that cause exit. For instance, none of the stimuli for driving the VCR are appropriate until after the **System On** stimulus, and they cease to be applicable after the **System Standby** stimulus.

Table 5.1 presents the entity/stimulus/response (ESR) lists for the VCR and it can be seen that the **Tape Eject** stimulus is the complement of the **Tape Presented**. Similarly, the **Stop** stimulus is only appropriate after one or more of the **Play, Forward, Rewind** or **Record** stimuli. These dependencies are marked by further indentation and the labelling of the associated stimuli with a numeric label, as shown in Table 5.1.

The initial purpose of the ESR lists is to prompt ideas regarding the state machine expression of system behaviour, but they later play another role in assisting the search for suitable objects to be used in the modularisation of the system. They also serve as checklists to be applied at various stages of the development to ensure that requirements are not overlooked. Also, it is sometimes useful to annotate entries in the ESR lists with known facts that emerge about the entities, stimuli and responses that might be helpful to the process of adding detail to the models.

Guidelines for the External View

Figure 5.2 is the External View for the VCR. Obviously there is no prob-

TABLE 5.1 Entity, stimulus, response lists for the VCR.

Entity List	Stimulus List	Response List
Television	System On	Forward
Aerial	--Change Channel	Reverse
User	--Modify Recording Program	Fast Forward
Tape	--Set Video Tuning	Stop Transport
Tape Drive	--Set Time	Play Head In
Channel	--Activate Auto Record	Play Head Out
Programme	----Deactivate Auto Record	Record Head In
Clock	--Tape Presented	Record Head Out
Frequency Band	----Tape Eject	Stop Head Operation
UHF Signal	--Play :1	
	--Forward :1	
	--Rewind :1	
	--Record :1	
	----Stop :1	
	--Auto Record Start	
	----Auto Record Stop	
	System Standby	

lem in following the syntactic rule that the complete system is to be represented by a single object, as shown, connected to the external objects that are of relevance to the system. However, the careful choice of externals, and the system's lines of communication with them, are crucial, which is where help is needed from guidelines.

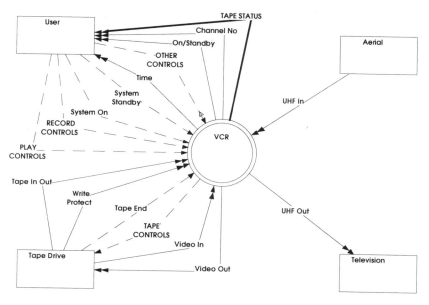

FIG. 5.2 The External View of the VCR.

Given ESR lists as a source of prompts, the next step in creating the External Model is to choose the entities to be shown as the externals on the External View. The required entities will be those with which the system needs to communicate directly. In the case of some products, there may be a requirement for a predefined type of connection to external systems, in which case the precise nature of the connection interface should be modelled. This clearly identifies the requirement of meeting the interface specification as a system responsibility. Pre-defined interfaces with externals may be of any form, including communications protocols, standard buses, or even plugs and sockets. In the case of the VCR, the system must interface with the standard broadcast Television signal, in the form of UHF, with connections to the aerial and television externals.

In contrast, when a system provides a service to human users, the user should appear as an external and an abstract form of interface communication, which specifies only the logical needs, as appropriate. The interpretation of the need and the provision of appropriate interface technology then becomes the responsibility of the system developers. Events and Parameterised Events are usually the most flexible and functionally precise means of representing the communication that a user wishes to have with a system. When a single command is to be relayed from the user to the system, an Event is used. For example, the VCR model has **Play**, **Record** and **Stop** events that are components of the **PLAY CONTROLS** and **RECORD CONTROLS** bundles in Fig. 5.2.

When more complicated information is to be routed to the system, a Parameterised Event is used, which indicates the desire for action, and also carries the associated parameters. For instance, when setting the VCR's clock, the user will have to supply the new time value. In the implemented system this action will require some dialogue between the user and the system, with the new time value built up in a series of steps. However, the precise nature of the dialogue is dependent on the physical interface between the system and the user. Decisions about this interface are delayed until the Transformational Codesign stage, as specifying them when developing the initial system model would make unacceptable commitments to the physical structure of the interface, and these are best made when the assignment of objects to software and hardware is made at a later stage. Moreover, the initial model can be used as the architectural specification of a range of products and the actual implementations of the products might achieve the interface with the user in a number of ways. For instance, setting the clock may either be achieved through a switch interface on the VCR, remotely, or through a more novel form of input such as speech.

Thus, in the VCR model, a parameterised event (**New Time**) is used to set the clock. It is held in the **OTHER CONTROLS** bundle in Fig. 5.2. Similar considerations apply to the information produced by the system for the user consumption, which is best shown as a Time-Continuous

Information flow. Output to the **User** shown on Fig. 5.2 is connected in the suggested style.

A high-level logical interface is also appropriate when the interface is with another computer system and its design has to be decided as part of a new product development, even though it might be known that a particular physical bus connection will eventually be used. The high-level interface presents the logical commands that flow between the two systems and is most generally modelled by interactions or events. An external interface that defines only the logical needs requires a decomposition that includes an object that services these logical needs. The way that the system under development and the external system are to connect and interact is therefore firmly placed within the scope of the new product development.

Tape is the remaining application domain entity in Table 5.1 which is of direct relevance to the system under development. However, it is not shown as an external in Fig. 5.2 since to do so would imply that a tape drive mechanism has to be developed for the product and this contradicts a specified Design Decision. Hence it is the specified **Tape Drive** that becomes the external, ensuring that only the interface needed to control the given tape drive will be engineered as part of the VCR development.

Guidelines for Model Dictionary Entries

The Model Dictionary defines the meaning of the labels placed on the connecting lines in a MOOSE model. For readability reasons, the labels on all the lines of communication in a model, but especially those between the system and the externals, should have been given carefully chosen names that imply their purpose. Even so, an External View will not be explicit enough for external review purposes, and it is certainly not suitable for its role at the top of the OID hierarchy, until dictionary definitions have been given for all the labels used on the connecting lines.

Table 5.2 gives the required definitions for the VCR. In general, some labels may be described only by comments at the start of the Behavioural Modelling phase, and these must be refined into definitive specifications before the end of the phase. However, the implied delegation of authority to supply the detail must be clear, and there must be no ambiguity with respect to intent.

Normally the definitions should only specify the logical structure of the connection, although some connections might contain implementation-level definitions in circumstances where there is no opportunity for flexibility. For example, the Television output and Aerial input in the VCR are both demodulated UHF signals connected to the system by coaxial cable. As an example of flexibility, the set of outputs to the **User** defined as **TAPE STATUS** consists of information flows for each aspect of tape status that is to be conveyed, but this only makes a commitment on information content

TABLE 5.2 Model dictionary for the VCR.

Connection	Entry
Channel No	channel /* number of channel */
On/Standby	'On | Standby' /* On\Standby indicator */
OTHER CONTROLS	New Time + Change Channel + Modify Recording Specification + Set Video Channel Tuning /* Setting Commands */
PLAY CONTROLS	Eject + Forward + Pause + Play + Rewind + Stop
RECORD CONTROLS	Record + Stop + AUTO RECORD
System On	'User generated on event'
System Standby	'User generated off event'
TAPE CONTROLS	Eject Tape + Stop Tape + Slow Forward + Fast Forward + Fast Rewind + Set Play + Clear Play + Set Record + Clear Record + Pause Tape + UnPause Tape /* Commands to manipulate the tape */
Tape End	'Tape end reached'
Tape In Out	'In | Out' /* Tape presence indication */
TAPE STATUS	Paused + Playing + Recording + Forwarding + Rewinding + Stopped + Tape Present /* Indication to the user about the status of the tape */
UHF In	'UHF signal from the aerial'
UHF Out	'UHF signal to the television'
Video In	'Video signal from tape drive'
Video Out	'Video signal to tape drive'
Write Protect	'ReadWrite | ReadOnly' /* Indicates the write permission of the tape */

and leaves the system developers free to decide on how to convey it. The choices available might include, for example, displays, indicator lamps or audible signals. Similarly, the main user input **OTHER CONTROLS** consists of four parameterised events, which are used to change the time displayed by the clock, change the current channel, set up the automatic recording facility, and tune the video to particular UHF frequencies. Options on how these might be implemented range from dedicated keys through to a general-purpose keypad, remote control or even voice input. As indicated above, the resultant system behaviour should be as specified, independently of how such a command is conveyed.

Guidelines for the Functional View

After the External View is completed, the next step towards completing the External Model is to produce a Functional View. As discussed in Chapter 4, this is a diagram that shows the major states into which the system can be 'driven'. As such, it partitions the total functionality of the system in the temporal dimension and this eases the problems of checking that the proposed functionality is as intended and complete. Moreover it brings out a 'feel' for system usability, since the major transitions that have to be triggered by the user to get to any particular state will be explicit, as will mutually exclusive aspects of functionality.

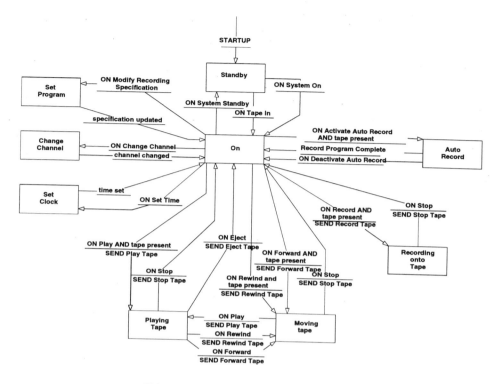

FIG. 5.3 The Functional View of the VCR.

The states shown on this diagram need not be limited to those entered as a result of external stimuli; for instance, a state such as *initialising*, which produces externally noticeable features such as delays, might also be included. It follows that some of the transitions might be due to internally detected conditions, such as 'initialisation complete'.

An approved Functional View plays a positive role in the construction of the Behavioural Model, as described in the next section.

Another guideline to be followed in constructing the Functional View is that states should be chosen that will be meaningful to the system's sponsors, not just its implementors, and they should be described in the terminology of the application domain. Figure 5.3 is a Functional View of the VCR. It is expressed in the modified STD notation described in Chapter 4, but other notations for defining state machines, such as Statecharts (Harel 1987), would be equally acceptable since they are for human consumption only and none of the MOOSEbench tools extract semantic meaning from this type of diagram.

For similar reasons, it is left to the discretion of the developers to decide where to terminate the explicit representation of a state and its associated transitions. Each state shown on the Functional View is potentially a state machine in its own right but, provided that the functionality to be offered while the system is in a state can be documented unambiguously by text, it need not be expressed in graphic form. Moreover, the diagram may not explicitly indicate the levels of concurrency in the system. For example, a programme can be set (**Set Programme**) while a tape is playing (**Playing Tape**), but not while the clock is being set (**Set Clock**). Again, text associated with the states can be used to describe the concurrency limitations. Alternatively, a more precise model of concurrency issues could be produced using the concurrent STD notation introduced in Chapter 4.

A number of checks should be carried out when the Functional View is considered complete. For example, every stimulus on the ESR lists must be shown at least once and, if a given stimulus has a role in several states, they must all have received mention. Similar checks should also be applied to responses.

5.3.2 Constructing the OID Hierarchy

The primary issue to be addressed by the guidelines for producing a Behavioural Model is how to ensure that the model will provide the appropriate architectural basis for the product development. We have argued that the role of a Behavioural Model is to identify the objects from which the system is to be constructed and the way that they will communicate and interact. As will be seen in later chapters, it is the Behavioural Model that is developed incrementally, first into an Executable Model and then through Transformational Codesign into an implementation of the product.

To produce a satisfactory uncommitted context for codesign, the Behavioural Model needs to be created in a style that maximises the range of implementation choices available. The objects need to have their functional responsibilities defined in an abstract manner that does not force a particular implementation style. At the same time, a distribution of functional responsibility should be sought that minimises the communication traffic.

Decisions concerning communication detail, such as the multiplexing of communication paths, should also be left open, which means that point-to-point communication should be used between objects, and intermediate objects should not be introduced as 'route controllers' if they have no other functional role to play. These policies also provide the benefit that the Behavioural Model can be treated as a model for a range or series of products and, even if a product range is not planned, the components of a model constructed in this style have the attraction of being well suited to re-use in other products.

Further to the aim of building flexibility into our models, a policy has to be established with respect to concurrency. The OID notation has substantial potential for concurrency, and some guidance is appropriate concerning the way in which this potential is to be utilised. In the authors' view a designer should always make full use of the concurrency potential in a model, to avoid unnecessarily limiting the concurrency that may subsequently be exploited in any particular implementation of the model.

This policy is clearly vital in parts of the system where performance reasons lead to the use of hardware, since hardware is intrinsically con-current. The policy is also necessary if flexibility is to be provided for selective placement of modest amounts of concurrency; for example, by multi-processing, or if the system is to be physically distributed as this enforces concurrency. It will be seen in later chapters that unwanted con-currency can easily be removed (or, perhaps more accurately, not imple-mented); certainly, adding new concurrency would be more difficult if the opposite policy were adopted and an essentially serial model was constructed.

Another issue that arises with respect to the design of the Behavioural Model is that experience has shown that, if different individuals with similar background and skills are given the definition of a notation and set a modelling task, almost without exception they produce structurally different models. Experience has also shown that this is not exclusively a MOOSE problem, and the authors have encountered the same difficulty with all the methods that they have tried. However, it is a particularly serious problem in MOOSE because of the aim to produce close to opti-mum designs for products that are destined to enter a competitive mar-ket. Intuitively it is difficult to accept that any one of a set of alternative Behavioural Models fits the architecture role equally well.

While our guidelines must take account of all the issues raised and lead to an appropriate modularisation of the system, it must be recognised that to achieve an appropriate and complete model may require some itera-tion. The 'appropriate' modularisation has to produce objects containing cohesive subsets of the total functionality, and different aspects of func-tionality have to be assigned to different objects. This separation of func-tional concerns is a major issue in design, and one on which our guidelines must provide strong assistance. However, they must also stabilise the model during the unavoidable iterative development. A *stable* model is one in which no significant structural changes are needed when essential changes are made to correct defects or add new functionality.

Of course, whatever guidelines we adopt will need to be applied with some subtlety, particularly in view of the hierarchical nature of the model development as this is concerned with the aggregation, rather than sep-aration, of functionality at the higher levels. Furthermore, in recognition of the fact that most difficulties with Behavioural Models arise out of human failure to interpret the structural and dynamic implications of the models correctly, attention to clarity is of extreme importance in the expression of object behaviour. In other words, a good model should contain components (or objects) with interfaces that present clear abstractions of their purpose and 'black box' encapsulation of their internal operation.

In summary, procedural guidelines are most important for the construction of the Behavioural Model, particularly in the stages of producing the highest one or two levels of the 'part of' hierarchy represented by OIDs. Since the definition of the high-level objects sets the terms of reference for their expansion at a lower level, starting a model on a sound and stable basis is vital. A basic aim in any form of top-down design should be to pass on structures that solve, rather than delegate, the difficult system problems so that, as the lower levels are approached, the designers' tasks get easier and the results become more predictable. Thus the authors propose guidelines to supplement the general advice given above, and recommend that the development of the Behavioural Model follows the folowing steps.

- Choose objects to represent application domain entities.
- Form a 'part of' hierarchy.
- Take additional prompts from the Functional View.
- Look for concurrency constraints.
- Apply functional decomposition.
- Address residual time dependencies.
- Terminate the hierarchy.

Guidelines for Entity Prompted Objects

Clearly the application domain entities in the ESR lists, which have been selected on the grounds of their relevance to the system, are candidates for becoming objects in the model. In fact such entities may prompt two types of objects: those that model them are called *real world objects*, and those that represent the system's policies with respect to their treatment we term *policy objects*. A real world object provides the system's communication with the entity and encapsulates the system's knowledge of its state. A policy object is an object that encapsulates some cohesive elements of the system's functional behaviour relative to one or more real world objects. This approach is a major factor in making the design process more deterministic and giving the model stability.

Some examples of objects prompted by application domain entities can be seen in Fig. 5.5, which shows the object **Channel Tuner**. The real world object **Chan** is intended to represent an abstraction of a television channel. The policy object **SigRou** is designed to present an abstraction of the strategy for handling and controlling channels in the VCR, such as routing the UHF signal to the Television when the VCR is inactive. The two other objects, **Mod** and **Demod**, are prompted by other guidelines given below.

An example of what is meant by stability of the model is that, if the operating policy of the system was changed, the **SigRou** object could be changed independently of the **Chan** object. Similarly, a different mechanism for setting the channel can be introduced into the **Chan** object without there being repercussions on the rest of the system. Moreover, as the **Chan** object is designed to have no application specific knowledge, it may be re-used in other systems.

Guidelines for Forming Part of Hierarchies

The danger in using the application domain entities to prompt the objects in the model is that there will be too many of them, and this would result in an unacceptably large number of objects appearing on a single diagram. This is where consideration has to be given to forming a 'part of' hierarchy so that, at higher levels, objects might be introduced to represent 'subassemblies' of objects, which are to be expanded into interconnected systems of simpler objects on lower-level diagrams.

This tactic is in some ways similar to the middle-out technique for creating a hierarchy of DFDs, as discussed in Chapter 3, but the clustering in our case is decided by inspection of the entities in ESR lists rather than a very large number of processes on a DFD. In fact, in inspecting the ESR lists, we are looking for the 'larger conceptual chunks' of the application domain. This is a very different process to aggregating 'bubbles' on a diagram. The aim in applying this step is to avoid the need to deal initially with the clusters rather than the members of all clusters, and later to consider the entities in each cluster separately from those of other clusters.

Guidelines for Using the Functional View

In using the entity-prompted approach to finding objects, some required functionality may be overlooked. A safeguard against this happening, and a way of finding 'large conceptual chunks', such as are needed in the hierarchy, are provided by the Functional View. A well constructed Functional View should mention every stimulus and response which enters and leaves the system. It also identifies the principal states of the system and shows them explicitly. Each of these principal states will be associated with a set of functions that normally have good functional cohesion, hence there is a case for assigning this 'cluster' of functionality to a single high-level object.

For each state shown by the Functional View, the responsibilities of the objects already in the model should be examined to establish whether or not they cover the functionality required by the state. Where functionality is not covered, either it can be added to existing objects, or new objects can be added to provide it. Considerations, such as the 'crispness' of the abstractions provided by the objects, will decide which alternative

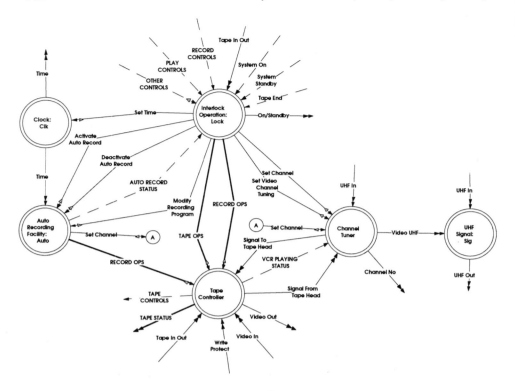

FIG. 5.4 The first-level refinement of the VCR.

method is chosen to add the new functionality. In the case of Fig. 5.4, which is the first-level expansion of the VCR system, the **Auto** object might exist because of a prompt from the entity 'Programme' in the ESR lists, or because the facility to programme recordings is specified as part of the Functional View in Fig. 5.3.

Guidelines for Finding Concurrency Constraints

Our primary goal so far has been to maximise the concurrency in the model. However, the concurrency may not be required if it appears at the interface of the system, as some aspects of functionality may be mutually exclusive for good usability reasons. Constraints of this nature should be specified by the Functional View, either as a result of the mutually exclusive behaviour being the responsibility of mutually exclusive states on the STD, or it may be stated in the textual description of the behaviour in a particular state.

For example, Fig. 5.3 indicates that the user cannot set the clock and the recording programme simultaneously. One strategy for modelling such a concurrency constraint in the Behavioural Model is to make the real world objects (or their policy objects) responsible for interlocking with each other. However, this has the disadvantage of distributing the policy with respect to concurrency constraints. It also results in a highly coupled system, which can make future change difficult and limit re-use. Therefore it is better to encapsulate the concurrency constraints in an object dedicated to the task of enforcing them.

This policy has resulted in the introduction of the **Lock** object in Fig. 5.4, which is responsible for determining the amount of concurrency presented to the user. It must be stressed that this object is only responsible for controlling the concurrency and is not an interface object with the user. In a similar way, the concurrency rules that govern the operation of the tape drive are encapsulated in an object that appears in the decomposition of the Tape Controller object.

Guidelines for Functional Decomposition

If the suggested guidelines have been followed, all aspects of required behaviour should now be covered by the model, at least at higher levels of the hierarchy. In principle, the guidelines can be applied recursively to all objects that require further decomposition, but another line of thinking that is useful in deciding how to partition objects into simpler objects is the 'responsibility driven' approach, as described by Wirfs-Brock and Wilkerson (1989).

Here the idea is to introduce objects that have specialised responsibilities and skills and that provide a service to other objects. As before, the objective is to encapsulate coherent functionality that represents a 'good' behavioural abstraction. In the VCR model, examples of these kinds of objects can be found in the 'Channel Tuner' object shown in Fig. 5.5. Here the **Mod** and **Demod** objects are made responsible for the modulation and demodulation of incoming and outgoing UHF signals.

The assignment of clear and coherent responsibilities to individual objects is of course paramount at all stages of model building, but to use this as the basis of making the top-level decisions on structure would take us into one of the most difficult aspects of top-down design.

Guidelines for Dealing with Time

One important issue concerning the behaviour of objects has not yet been addressed, namely that of how objects whose behaviour is dependent upon time obtain the relevant timing information. In general, there are two types of time information that an object might require; either it needs periodic timing signals in order to fulfil its behavioural responsibilities, or it requires absolute time information; e.g., time-of-day information in the form of day, hour and minute. The VCR model has an example of the latter with the auto recording object. An example of the former can be found

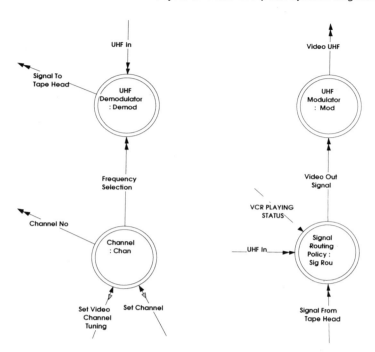

FIG. 5.5 Decomposition of the channel tuner object.

in the Mine Pump system model of Chapter 4, where the **Log** object periodically receives a time event on which it bases the majority of its actions.

The question arises, how should time be incorporated into the Behavioural Model? One temptation is to list time as an entity and then to use it to prompt a timer or clock object in the model. Indeed, it may appear at first sight that this approach has been used in the VCR model, since the entity 'clock' appears in the entity list. However, this is rather an unusual case. It appears in the entity list primarily because the system's interface presents the user with the time, and the user can carry out operations directly on the clock. The time of day is therefore a functional aspect of the VCR. If the time were not directly presented to the user, it is unlikely that this entity would appear and the requirement for the **Auto** object to know the current time would not be satisfied unless other means were used. (A further example can be found in the Mine Pump model in Appendix 2, where the **Log** needs to be informed of periodic events but no clock or timer appears in the entity list.)

The general policy of adding time to the entity list is not adopted, for two reasons. First, it is more conventional in OO to regard time as an attribute (a property of an object) rather than as an entity. For example, Booch (1994) states: '...attributes such as time, beauty or color are not objects, nor are emotions such as love or anger.'

This view is compatible with regarding timing behaviour as a non-functional attribute of the system. However, we must acknowledge that in many computer systems we must include an object that makes time available to other objects. This brings us to the second reason why time is not treated as an entity, namely the requirements of this object in terms of the resolution of the time it produces and whether it produces absolute time or periodic events. These are best determined when the other objects have been chosen and their functional responsibilities identified. Thus the MOOSE policy with respect to this issue is to identify the objects in the model and then to determine their timing requirements. Objects that need timing information can then have their needs satisfied by the introduction of a timer object. In the VCR model, the timer object is within the **Clk** object. In the Mine Pump model, it appears because of the requirements of the **Log** object.

Guidelines for Terminating the Hierarchy

The refinement of the Behavioural Model leads to the hierarchy of OIDs, which stops when only primitive objects remain. It is therefore important to system designers to know whether or not an object should be treated as primitive. The key issue in making a decision is to consider the strength of cohesion that applies to the responsibilities assigned to the object. This can partly be achieved by reviewing the input connections into an object and thinking about the consequential behaviour of the object. If the actions required of the object exhibit a high degree of functional cohesion (that is, the inputs all contribute to a single, well defined operation) it is likely that treating it as primitive is a good choice.

One informal way to decide if an object should be primitive is to consider the nature of the STDs that would be required to describe its operation. If it is clear that the object would have only a single well defined state at all points in time, then it should be primitive. If the description of the object's behaviour would require more than one state machine or a single state machine that has many states, the object is probably too complex to be primitive.

A similar, but slightly different, approach is to consider the internal structure of the class definition for the object. In MOOSE, a Class Implementation Diagram (CID) would be drawn, the structure of which will be discussed later. Here it is sufficient to know that it shows an object's encapsulated state and the functions that operate on this encapsulated state. If there is a high degree of internal coupling – that is, functions use other functions or they operate on the same set of internal state information – the object is primitive. It should be noted that a primitive object need not be one that sustains only a single thread of control. Indeed, primitive objects may well have internal concurrency but, as before, there must be a reason for the concurrency being confined to a single object.

Considering the VCR model, the **Tape Controller** and the **Channel Tuner** shown in Fig. 5.4 are shown as non-primitive objects, primarily because they are composites of lower-level objects that were identified from the ESR analysis and have been gathered together for readability. The other objects in the model are primitive.

Guidelines Summary

In summary, the guidelines for producing a MOOSE object model assist in partitioning the required functionality on an external entity, a temporal or a specialist basis, and this is continued until primitive objects are reached. Of course, careful review of the result is desirable. What helps at this stage is the fact that, as with most things that humans find hard to do, they can usually tell when it has been done well. Good Behavioural Models should allow functionality to be inferred easily from the diagrams without extensive need to refer to textual explanations. The purpose of the interactions in the diagrams should be clear, and the level of detail at each level of the hierarchy should be sufficient to prompt understanding without being distracting. We hope that the example used in this chapter displays these characteristics.

5.4 CONSTRUCTING THE DOMAIN MODEL

To complete the architecture as indicated by Fig. 5.1, a Domain Model also has to be developed. Its role is to capture application domain knowledge for use by system developers later in the design process. The main relevance of the Domain Model will be seen in Chapter 6, where class definitions are developed for the objects in the Behavioural Model.

A number of software engineering paradigms have concepts similar to the Domain Model. For example, the Initial Object Model of OMT (Rumbaugh *et al.* 1991) is described as showing 'the static data structure of the real-world object classes and their relationship to each other'. As far as the MOOSE paradigm is concerned, this is quite satisfactory for representing the domain information ,and so the OMT notation has been adopted for this role.

A Domain Model of the VCR is presented in Fig. 5.6. This has been constructed by following the guidelines in Rumbaugh *et al.* (1991), an outline of which was provided in Chapter 3. It uses the technique of identifying problem specification nouns and verbs in the product proposal (presented earlier) to assist in identifying the high-level objects.

The OMT method provides guidance with respect to the level of abstraction of the problem statement, in particular suggesting that details of computer implementation technology and system internals should be avoided. It also provides heuristics for identifying the key

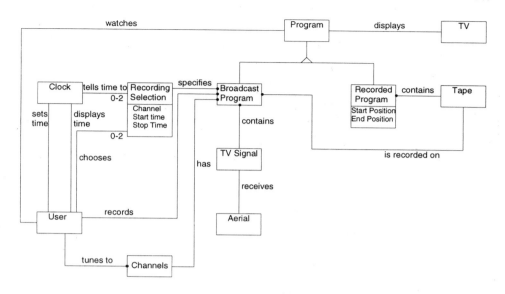

FIG. 5.6 The domain model for the VCR.

classes of relevance to the system, the associations between classes and the attributes of classes. This will clearly lead to a model at a very high level of abstraction, in terms of the real world entities that interact with the proposed system but which are not part of its structure. Hence classes such as 'Tape Drive', 'Tape Heads', etc., do not appear in the model.

The 'Clock' class, which is shown in Fig. 5.6, could perhaps be considered to be an internal component of the system. However, given that the user of the VCR will be able to set the time of the clock and to see the time displayed, it was considered to be appropriate to include it in the Domain Model.

Given the guidelines on the level of abstraction of the problem statement, it is not surprising that many of the classes in the OMT Model appear as externals in the MOOSE External Model of the VCR (Fig. 5.2), albeit with a rather different interpretation. Most of the classes and associations should be self-explanatory; however, the classes 'Recording Selection' and 'Channels' probably require some comment. 'Recording Selection' represents the programs that the user selects for automatic recording, and 'Channels' portrays the UHF frequencies to which the user tunes the VCR after purchase.

The simplicity of the model in this case is probably due to the fact that the OMT approach was developed from data modelling techniques widely used in information system development. Most of the stored data in an information system will relate to application domain entities and, for anything other than simple systems, the relationship between these entities is likely to be highly complex. Thus an OMT model, constructed according to the guidelines, will be very useful in clarifying these relationships and providing a structure for the system. The approach, however, has its limitations. For embedded systems such as the VCR, significant complexity lies in inventing abstractions that provide the behaviour required of the system – in other words, in defining the internal structure and operations of the system. In this context the OMT guidelines are very limited.

5.5 SUMMARY

This chapter has addressed the issue of developing an Architectural Model that forms the basis of future product development, the Model having first been approved by the client. The Architecture comprises a Classified Requirements Document, together with a Behavioural Model and a Domain Model, and guidelines have been provided on how to develop these.

Each of these components of the Architecture are used in later stages of the MOOSE paradigm in developing a more detailed product specification and design. In particular, Chapter 6 addresses the issues of developing the Behavioural Model to allow it to be executed, and thereby providing a deeper understanding of its behavioural characteristics.

6 Creating an Executable Model of a Computer System

Computer systems have a level of internal activity that exceeds human comprehension. Modern processors are capable of executing many millions of arithmetical and logical operations per second, and special purpose hardware performs operations even faster. Clearly a static 'paper model' that defines the behaviour of such a system is a very inadequate representation of the system's capabilities, features and possible failings. Only 'hands on' contact with a working model can begin to offer a satisfactory means for the evaluation and validation of specified behaviour. Thus the best way to evaluate a computer system model is to simulate the operation of the system that it models, and to conduct planned experiments using the simulations.

If a model fully specifies a system's responses to its inputs, a simulation of external behaviour can be created in which the user provides the inputs and receives the associated responses. If the model also specifies the internal operation of a system, as a MOOSE model does, a simulation can allow users to observe the propagation of activity through the various components of the model. At a later stage of development, when implementation decisions are being taken, timing estimates for the operations within the components of the system and the propagation delays on the communication paths between them can be added to the model, thereby allowing valuable performance and timing estimates to be derived from the simulation.

This chapter describes an extension to the notation for behavioural modelling that was presented in Chapter 4, which allows sufficient detail to be added for the specified behaviour to be simulated. Figure 6.1, which summarises the MOOSE paradigm, shows that the addition of this detail takes place in the phase that follows the acceptance of the Product Architecture, and it produces an Executable Model. The immediate purpose of this model is to provide a means to gain approval for the behavioural aspects of the architecture. In effect, the development of the Executable Model provides a working prototype, which facilitates experiments aimed at reaching agreement on the specified external behaviour. Of course, at this stage of development, the simulation will demonstrate only

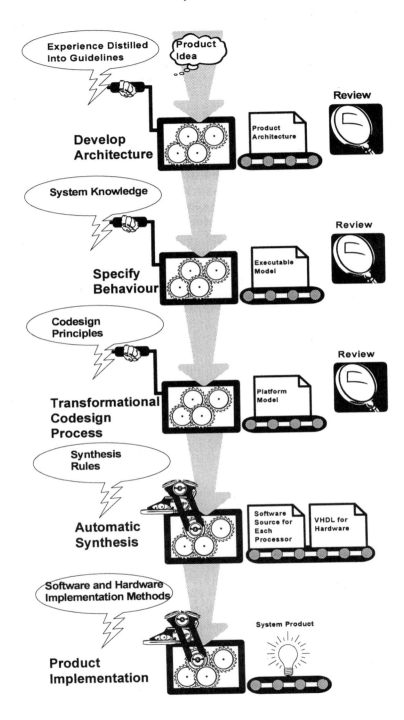

FIG. 6.1 The MOOSE Paradigm.

logical behaviour with respect to inputs and outputs. It does not model the style of the interface, since the specification of this kind of detail is not included in the model until the Transformational Codesign phase.

A second use of the Executable Model is to provide a means to validate the correctness of its internal structure and operation. It is important to establish that this structure is sound, since it is carried forward into an implementation of the product during the Transformational Codesign phase. For example, the robustness and safety of operation of the design decisions embodied in the model with respect to the relative timing of external stimuli need to be explored, and this can be done by experiments that use the Executable Model. Clearly, absolute timing issues cannot be explored at this stage since, by policy, the model does not contain the implementation detail on which they would depend. Later in the development process, after specific commitments have been made to component technology, the timing and performance experiments can be conducted. This is described in Chapter 9.

It follows from common sense and good practice that, if an Executable Model has been subjected to rigorous validation and timing studies, its detailed behavioural specification and internal structure should be carried through accurately into implementations of the architecture. The technique described below for making an Executable Model involves the addition of source code for the operations of its primitive objects. In the Automatic Synthesis phase of the MOOSE paradigm, this same code is combined with other information in the model to produce actual implementation source code for the product.

Natural questions arise from the concept of a model that both can simulate the behaviour of a product and from which implementation source code can be synthesised. These include: 'What is the relationship between an executable model and an implementation?'; 'How does an executable model of a system differ from an implementation?'; and 'What is the cost of constructing an Executable Model in comparison with the cost of implementing the system?'. The answers to all these questions, which concern the economic justification for the effort of making a model executable, are considered in Section 6.3. First, a basis for the further discussion of executable models is set by Sections 6.1 and 6.2, which describe how the Behavioural Model for the VCR architecture described in Chapter 5 is rendered executable. Later sections define the dynamics of such executable models and how these dynamics can be simulated. Examples of the uses of the Executable Model are discussed in Section 6.6.

6.1 CREATING AN EXECUTABLE MODEL

The notation introduced in Chapter 4 allows a behavioural specification of a computer system to be expressed as a hierarchy of Object Interaction

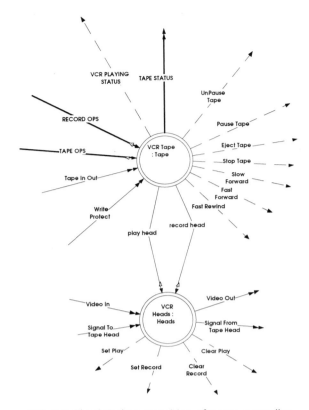

FIG. 6.2 The OID decomposition of a tape controller.

Diagrams (OIDs), such as that shown in Fig. 6.2, terminating at primitive objects, which have textual descriptions called Object Specifications (OSPECs). Although these diagrams represent a hierarchical 'part of' decomposition of the system, for the purpose of understanding behaviour they can be regarded as equivalent to the 'flat network' of primitive objects that would result from replacing each non-primitive object by the interconnected set of objects shown on its OID. Thus only the primitive objects need to be considered when the detail to support execution is added to the model. Non-primitive objects in a model make no independent contribution to system behaviour, although they play an important partitioning role, encapsulating the interworking detail of their component objects and presenting an interface that is the externally required subset of the interfaces of these component objects.

In general there are two contrasting approaches that can be applied to the production of an Executable Model. One is to provide a software environment that builds its own internal representation of a model, and then *interprets* the semantics of the model in order to produce equivalent behaviour. The alternative approach is to produce software that *synthe-*

sises a program from the information contained in the model, which can be compiled and executed to provide directly the required behaviour. In both cases the semantic detail concerning objects and the way in which they interact can be obtained from the graphical part of the Behavioural Model, and additional definitions (in programming language style) are needed as specifications of the operations of the primitive objects. Thus, in both cases, the effort required from the model builder is similar. However, the second approach has been chosen in MOOSE because an Executable Model produced as a stand-alone program can be tested, exercised and controlled in a standard program development environment. Also, in general, the execution of a compiled model should be faster than an interpreted model, and hence more extensive experimentation becomes feasible.

The programming language chosen for the synthesised simulation software is C++, arguably the most widely used object oriented programming language in industry. In fact, at the level at which code for the basic operations has to be provided, namely functions, C++ closely resembles C, which currently is even more widely used than C++ for the class of system developments for which MOOSE is intended.

In summary, the conversion of a Behavioural Model into an Executable Model requires only the addition of information from which the class definitions for its primitive objects can be constructed automatically, which in practice amounts to less effort than would be required to write the complete class definitions in C++. Given these class definitions, the rest of the program which deals with the instantiation of objects and simulation of the communications between them is synthesised from the information implicit in the graphical part of model. The user-provided C++ code interfaces to the synthesised communication mechanisms through calls to 'built-in' functions, as described in Section 6.2.3. Therefore the authors believe that the synthesis approach used by MOOSE minimises the amount of work that a model builder has to invest in making a Behavioural Model executable.

6.2 CREATING CLASS DEFINITIONS FOR PRIMITIVE OBJECTS

To satisfy the requirement for automatic construction of class definitions, three additions have to be made to the model for each class of primitive object. These are: a textual Class Interface Specification (CIS); a Class Implementation Diagram (CID); and textual specifications in C++ for each of the functions and data stores (FSPECs and DSPECs) shown on a CID.

A general point that applies to all three additions concerns the conventions to be used for names. In all cases, the names used for the functions and the input/output connections of objects become names in a synthesised C++ program, and the normal C++ conventions should be observed. Also, since object definitions are potentially re-useable, the names chosen for their interface connections should be appropriate to the abstractions they represent. However, the names already introduced as labels on the lines connected to objects in the Behavioural Model will have been chosen to provide good readability of the model and will follow application domain terminology. Therefore a mapping is needed between one set of names and the other, and this is appended to the OSPEC of each primitive object. The examples given below, which illustrate the development of the VCR model in Chapter 5 into an Executable Model, assume appropriate mappings exist. They are defined explicitly in Appendix 4, which contains the complete Executable Model for the VCR, and a fuller description of mapping information is given in Appendix 1.

6.2.1   Class Interface Specifications (CISs)

A CIS defines both the inputs and the outputs of the objects in a class, although only the inputs are required for the C++ class interface. The output specifications are provided because they are needed by the 'routing' objects included automatically in the simulation program, to ensure that outputs get to their destinations. Typically, the inputs will be the interactions supported by the class, the information flows that it uses, and the events to which it responds; the outputs from a class can be interactions with other objects, information flows or event signals. A straightforward format is used for the entries in a CIS, which consist of a keyword indicating a kind of connection, followed by definitions of actual connections and terminated with ';'. The permitted keywords for specifying connections are:

> INTERACTION IN
> INTERACTION OUT
> EVENT IN
> EVENT OUT
> PARAM EVENT IN
> DISC INFO OUT
> CONT INFO IN
> CONT INFO OUT

The last three keywords relate to Information Flows (discrete and continuous). Definitions of interactions give their name, parameter names and types, and result type. Information flow definitions specify a name and type structure, and event statements give the name of the event. If user-defined types are required, they may be defined by a type statement as in the example below, which is the CIS for the **VCR Heads** class of object used in the VCR model shown in Fig. 6.2.

```
CIS VCR_Heads
   INTERACTION IN: play void (BOOL);
   INTERACTION IN: record void (BOOL);
   EVENT OUT: playing;
   EVENT OUT: not_playing;
   EVENT OUT: recording;
   EVENT OUT: not_recording;
   CONT INFO IN: signal_in TVSignalType,
      from_heads TVSignalType;
   CONT INFO OUT: signal_out TVSignalType,
      to_heads TVSignalType;
   TYPE TVSignalType DEFINITION (int);
```

The full syntax of a CIS is given in Appendix 1, where it will be seen that there are some additional statements beyond those illustrated here – e.g., those for use with respect to dynamically created objects – which are described in Section 6.2.6.

6.2.2 Class Implementation Diagrams (CIDs)

Superficially, CIDs have a similar appearance to OIDs in that they contain 'bubbles' connected by lines. However the 'bubbles' on CIDs are drawn as a single circle, and each such 'bubble' represents a function, which might for example implement an operation (method) of the class. State information of the class is represented on a CID in the style of the 'data stores' used on DFDs; that is, as the name of the information or data placed between a pair of horizontal lines.

Normally (see Section 6.2.4 for the exceptions), all the interaction, information and event lines entering the object are shown connected to the functions that have the responsibility for dealing with them. This is illustrated in Fig. 6.3, which is the CID for the class **VCR Heads**, required by the **Heads** object in Fig. 6.2. A function may only have one incoming interaction or event, but it may have any number of incoming information flows and outgoing interactions and events. These are not arbitrary conditions; the restriction to a single input stimulus conforms to the familiar programming language convention that a function only has one entry point, whereas a function can make unlimited calls on other functions to access information flows and produce outputs.

A CID may also contain *private* functions, again depicted by 'bubbles', which are only used by other functions on the same CID; and several kinds of *special* functions, which are introduced later. Functions that use other functions are connected to them by interaction lines, and access to data stores from the functions is indicated by connecting them by time-discrete information flows. It might be helpful, when designing the CID

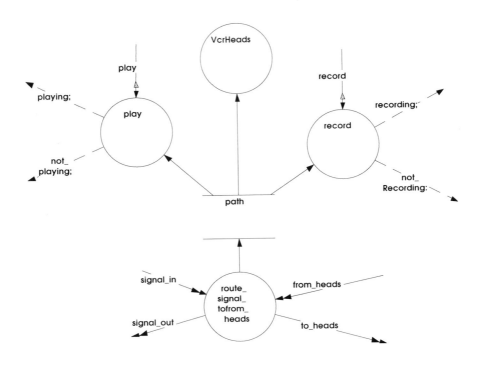

FIG. 6.3 The CID for **VCR heads**.

for a class and defining the functions and data that it depicts, to remember that in a compiled program all objects of a class will share the same copy of the function code but each object will have its own copy of the state information.

Definitions of the functions and data in a class (FSPECs and DSPECs) are given in textual form and this text is linked to each store and function symbol in the same way that OSPECs are linked to the symbols of primitive objects on an OID. In order to use the synthesis capabilities of MOOSEbench, these definitions must be written in C++ and be name- and type-compatible with the definitions given in the CIS. For example, the definition of the data store **path** is given by

```
Store Definition path
    typedef enum {PLAYING, RECORDING, NONE}
        PathType;
    PathType path;
```

and the specification of the function **play** is given by

```
void play (BOOL play) {
    if (play) {
        path = PLAYING;
        CAUSE (playing);
    }
    else {
        path = NONE;
        CAUSE (not_playing);
    }
}
```

The function **CAUSE** () is an example of a built-in system function (see Section 6.2.3) whose purpose is to produce an event signal on the outgoing event connection named by its parameter.

In summary, the connections into a CIS carry forward from the instances of the class (objects) that appear on OIDs, with connection labels modified according to the Mapping Specifications. The design of a CID concerns the choice of an interconnected set of functions and data to implement the required object behaviour, and this requires skills similar to those applied to the design of the OIDs. Choice of an initial group of functions to receive the interactions and events is straightforward and the significant design issues arise in choosing the supporting functions and possible classes (see Section 6.2.4) that are to be used. Rather more specialised programming knowledge will be needed to provide the necessary special functions described below. If appropriate diligence has been applied to preparing the Behavioural Model, the information in its OSPECs should make the task of coding the functions also straightforward. In cases where the data stores need to contain complex application-specific information, the Domain Model component of the Architecture should be helpful in determining the structures to be used.

Special Functions on CIDs

In addition to the method functions, event functions and private functions already introduced, there are four kinds of *special* functions that can appear on a CID, namely *constructor* functions, *destructor* functions, *state machine* functions and *active* functions. Readers familiar with C++ will be aware of the purpose of the first two, and the second two are MOOSE-specific concepts.

A *constructor* function is one that is executed automatically whenever an instance of the class is created. They are distinguished (as in C++) by having the same name as the class, and an example can be seen on the CID for the **VCR Heads** class (Fig. 6.2), where the constructor function **VCRHeads** is responsible for setting the data store **path** into the correct initial state. Constructor functions can have parameters, whose values are defined for each particular instantiation of an object in its OSPEC, where in fact the class name will also appear. Figure 6.3 provides no example of this, as all instantiations of the **VCRHeads** class require precisely the same action on initialisation. However, there is a good example in Appendix 3, where the sensors of the Mine Pump system are initiated with different parameters, depending on whether they are monitoring methane or carbon monoxide levels.

Destructor functions are also distinguished by using the C++ convention of giving them the same name as the class preceded by a '~' and they are called automatically when a dynamically created object is deleted. These functions are not required unless the design of a class makes use of other *unusual* features of C++ to allow one of its objects to have 'possessions', or cause disturbance outside of its encapsulated state. This is not encouraged in MOOSE. An example that provides a marginal justification is where a dynamically created object creates child objects that are to die when it dies, which is plausible in software but harder to visualise in hardware.

Active functions are particularly important in MOOSE because they are the means by which the concurrency is maintained in an Executable Model. We have previously said that objects should be regarded as concurrent, but we have not defined whether this means that they might be subjected to simultaneous external stimuli, hence several become busy at the same time, or that they have their own internal drive that keeps them permanently busy. In fact MOOSE provides both, in that events stimulate the operation of event functions and active functions operate permanently. Active functions are distinguished rather by default, as they neither have constructor/destructor names nor any interaction or event line entering them.

Typically, active functions might be continuously monitoring information flows or data stores that are liable to change as a result of external effects or actions in other functions. They are the only functions that make an object capable of independent action, since the stimulus for all other functions arises from explicit action outside the object; for example, an interaction request or the occurrence of an event. Figure 6.3 shows one active function, **route_signal_tofrom_heads**, whose function is to carry out the appropriate routing of the incoming time-continuous information flows **signal_in** and **from_heads**.

The final kind of special function, the *state machine* function, has a distinctive graphical representation, namely a circle with a broken outline (see Fig. 6.4). Such functions provide alternative ways of handling events, and they are allowed to receive any number of events and react by producing other events or by calling functions. However, in practice, state machines need to be controlled, and hence they also provide an interaction **SetState**, whose parameter allows the operation of the state machine to be *halted*, *started* or *reset* to its entry state. Thus a state

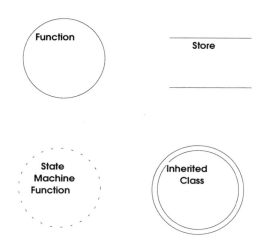

FIG. 6.4 CID symbols.

machine function is given the chance to act whenever one of its input events is caused or its **SetState** interaction is invoked.

The behaviour of a state machine function is described by an STD, similar to that used as a Functional View (see Chapter 4), and its input events provide the stimuli that cause transitions. The effect of the **SetState** function is the same in all instances and a formal definition is not required. The outputs may be events or interactions and these appear on the STD as the actions taken when the transitions occur.

State machines are provided in MOOSE for two reasons. First, there is an intrinsic need for them in control systems, as can be seen in the Mine Pump example in Appendix 3, and, although the same effect could be 'programmed' into the normal kinds of functions, it would be an inconvenient and probably error-prone technique. Second, when an object is implemented in hardware, any state machines that it contains will be translated easily and directly into hardware; they can of course be implemented in software but it is probably safest to adopt a stereotyped structure. Either form of implementation can be produced automatically from the STD representation.*

6.2.3 Built-in Functions for Use in FSPECs

Functions on a CID are defined by FSPECs that are coded in standard C++ (Stroustrup 1991), and no explanation of this will be given here. However, when the C++ code needs to access the input and output con-

---

* The facilities for interpreting STDs used in an Executable Model are not in MOOSEbench-1, therefore they have to be treated as active functions and are hand coded in C according to a template that is provided to simplify the task.

nections to a function, it should do so by calling functions that have *built-in* definitions supported by both the infrastructure supplied to an Executable Model and synthesis of implementation source code. The basic idea is that pseudo function calls (to the built-in functions) are written as if the functions in question were actually to be called, and the required effect will be produced by substituting appropriate C++ code.

For example, a function that has an outgoing interaction line connected to its symbol on a CID needs to use this interaction, but normally, at the time it is defined, the object that is to service the interaction will not be known. Such interactions have **INTERACTION OUT** entries in the CIS of the containing class, that will give its local name and parameter specifications. A built-in function CALL is provided to allow such interaction calls to be specified by the C++ statement,

```
CALL (local_name_of_interaction, parameters);
```

of course with the appropriate substitution of detail. The MOOSE synthesis mechanisms will produce an effect equivalent to the more familiar type of C++ statement,

```
destination_object_name.destination_interaction_
    name (parameters);
```

and, although this detail should not concern the user, the effect will normally be achieved by using a pointer to the destination object. Other built-in functions provide, for example, for reading and writing information flows and causing events; the complete list of available functions is:

CALL – Calls an interaction
GET – Reads from a Time-Continuous Information Flow
PUT – Writes to a Time-Continuous Information Flow
RETRIEVE – Reads from a Time-Discrete Information Flow
SEND – Writes to a Time-Discrete Information Flow
CAUSE – Sends a named event
DCALL and **CREATE** – See Section 6.1.5
TIME CALL – See Chapter 9

Examples of the use of most of these built-in functions can be found in the full VCR model in Appendix 4. Appendix 1 contains further details of their syntax and operation.

6.2.4 Introducing Class Hierarchies

For readers unfamiliar with OO programming, this section may be ignored, since its only relevance is to the exploitation of the more advanced concepts of OO that allow class definitions to make use of other class definitions. Other readers will be aware that traditional methods of OO analysis and design place great stress on developing these 'kind of' hierarchies into class definitions through the use of inheritance mechanisms. The MOOSE approach presented here has so far concentrated on objects and has ignored inheritance and the development of a 'part of' hierarchy. Of course we recognise a single-level 'kind of' relationship between an object and its class definition, but we have not considered extending this to defining classes in terms of other classes.

There is a good justification for the approach that has been taken, which arises out of the fact that we have been concerned with the architecture and dynamics of a 'machine', albeit an abstract machine that is manifest in the behaviour of a computer system. The authors believe that engineers who design machines think in terms of parts, and the way in which they will work together in the machine. Our concern has not been to 'model the world in objects' (Booch 1994), except marginally in producing a Domain Model, but rather to model the behaviour of the machine and specifically that of its component parts.

However, at the stage of defining classes for the primitive objects, we might observe that some of the parts of the machine (objects) share common features, and some have highly specialised facets of behaviour that could usefully be given their own class definitions. These aspects of class definitions are best dealt with by making use of inheritance. Therefore MOOSE provides for the development of a class hierarchy that allows the required class definitions for primitive objects to be organised in a more conventional OO style. Its use is strictly optional and some may feel that the technique is only worth consideration with respect to class definitions destined to be placed in libraries so that they may be re-used in similar applications at a later time.

The provision is based on two specific features of the notation and tools that support its use. Firstly, using MOOSEbench, any number of class definitions may be added to a model, regardless of whether or not they directly correspond to primitive objects in the model. Secondly, any CID of any class definition may specify the inheritance of any other class definition present in the model. Therefore, if it is thought desirable that the class definition for a particular primitive object should be constructed in part from two other class definitions, all three class definitions would be added to the model, and the one that is to used to instantiate the primitive object would be defined to inherit the features of the other two.

In fact it is desirable to provide two forms of inheritance, 'public' and 'private'. Public inheritance makes the interface (hence external behaviour) of the inherited class part of the interface (hence behaviour) of the inheriting class, and private inheritance makes only the functions of the inherited class available to those of the inheriting class. Both types of inheritance are shown on a CID by using the object symbol that is normally used on OIDs. We should think of this more generally as a class-

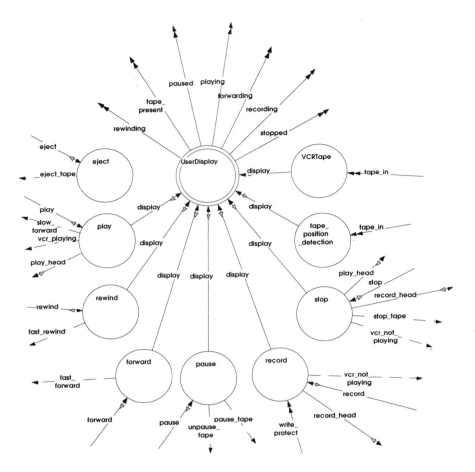

FIG. 6.5 The CID for **VCRtape**.

identifying symbol rather than an object symbol. When it appears on an OID, it represents an instance of the class, and when used in another class definition (i.e., a CID), it indicates inheritance. The distinction between private and public inheritance depends only on whether the use of the inherited class is restricted to other functions of the CID on which it appears.

An example of the use of inheritance is provided in the VCR model by Fig. 6.5, which is the CID for the class **VCRTape**. The inherited class, **UserDisplay**, is privately inherited, because the interface to this class does not contribute to the interface of the inheriting class.

Some uses for inheritance will normally emerge spontaneously as the designer is constructing class definitions for primitive objects. For instance, the need to repeat the same detail in several class definitions will become apparent, hence it will be natural to place the repeated detail in a separate class definition. Alternatively, through the Domain Analysis and Domain Model produced in the first phase of MOOSE, the designer should have discovered hierarchical relationships among the entities of the application domain that can now be used to prompt ideas for building inheritance into the class structures specified by the CIDs.

Looking beyond the immediate model, if a library of classes exists, some of its classes may nearly fit the requirements of a new model, but to make them fit exactly they have to be customised a little. The cleanest approach to this problem would be to create new classes, while inheriting the library classes in order to provide most of the implementation.

6.2.5 Imported Classes

We have assumed so far that all the required class definitions will be provided in the MOOSE notation. This is not strictly necessary since all the definitions in MOOSE notation are translated into C++ source code when either the Executable Model or the implementation source code is synthesised. Class definitions already written in C++ could equally well be imported.

Examples of where the need for this might occur are when part of a new development in MOOSE overlaps an existing system implemented in C++, or when, for various reasons, part of the new development is carried out using different methods and tools.

The notation for introducing class definitions already available as a C++ .src or .exe file is described in Appendix 1. These class definitions can only provide a restricted interface to the MOOSE classes in the form of **INTERACTION IN** connections, and they cannot themselves connect into MOOSE classes unless they have a direct knowledge of the C++ for the MOOSE model.

6.2.6 Dynamically Created Objects

One of the attractions of Object Oriented programming is that objects can be created and deleted dynamically. This is useful, for example, when a computer system has transient activities, which can be defined by a class definition that encapsulates the detail of the activity, and each instance of the activity has a clearly defined birth and death time. An object with a longer lifespan can then use the class definition to create (and later delete) dynamically an object of the specified class as and when the need arises. Although dynamically created objects do not map easily to the hardware of a computer system, they are often very useful to its software, and might even be relevant to some dynamically configurable hardware such as Field Programmable Gate Arrays (FPGAs), which can be reprogrammed during a system's operation..

During the lifetime of a dynamically created object, its creator can optionally use its interactions, but it must always interact with the constructor function at the time of the object's creation. Hence dynamically created objects can be recognised on the OIDs by the distinctive 'interact with constructor' lines that connect them to their creators. Other interactions with the operations of dynamically created objects use normal-style interaction lines. Also, if the creator provides a suitable interaction in its interface, pointers to the created object can be passed to other objects, which may then interact with the created object. Chapter 4 explained how dynamically created objects and their creators and connections are shown in an OID model. We now need to consider how to implement this feature of a model within the class definitions of the creator and other objects that are to handle pointers to a dynamically created object.

Dynamic objects are created by using the built-in function (see Section 6.2.3) called **CREATE**, which will have the effect of creating an object of the appropriate class, calling its constructor function and returning a pointer to it. Objects may then interact with the dynamically created object through the built-in function call **DCALL**, which is very similar to a standard function call (**CALL**) but with an additional parameter specifying a pointer to the object. Because dynamically created objects are almost certainly implemented in software, events and information flows are not permitted as inputs, although they may be produced as outputs. Objects are deleted through the standard C++ function **delete** (), which requires a pointer to the object as a parameter. The use of dynamic objects requires the careful specification of the TYPE of the class interfaces in the CIS. The full rules are presented in Appendix 1. Appendix 5 contains a completed model for the example of dynamic creation introduced in Chapter 4.

6.3 COMPARING AN EXECUTABLE MODEL TO AN IMPLEMENTATION

Any executable model should behave like the system that it models in some respect, but in some methods the models might exhibit limited and specialised views of a system, such as a state machine view or a performance view. The MOOSE Executable Model aims to simulate the full functional behaviour of the system. Thus it is appropriate to examine how this model, which behaves in the same way as a proposed product, compares with an implementation of the product. Clearly, if the difference is small, we need at least to be sure that the work done in producing the model can be used in the final product and need not be repeated. Furthermore, since the MOOSE Executable Model is to be used at a relatively early stage in the development, we need to show that it involves significantly less effort than the production of an implementation. Also, regardless of how close a model is to an implementation, we need to understand the nature of the differences since they represent the gap that has to be bridged in the codesign phase of the development.

In general, an Executable Model is a 'soft' model of a complete system in which the behaviour of primitive objects is defined in terms of the software functions that define the operations of the objects. If the model as a whole can provide a realistic operating context for each primitive object, and if the language of the model can be translated into an acceptable implementation language, then the model and implementation intersect for those objects that are to be implemented in software and hence there is no expenditure of 'throwaway effort' on these objects. Even though completion of a full Executable Model will require the production of class definitions for other objects that eventually become hardware, the added work in producing these 'soft emulators' of hardware objects is small in comparison to that of fully specifying their design, implementing them and integrating them into a system. There is also a strong justification for incurring the cost of producing 'soft emulators' of hardware objects, since their existence ensures that the other (software) objects of a system can be fully tested within the Executable Model, ahead of hardware production and without recourse to a 'breadboarded' hardware prototype that, in practice, can too often have an out of date or incomplete specification.

If the relatively small effort that is required to produce 'soft emulators' of hardware objects is felt to be a significant overhead on hardware development, then an execution environment for the models can be provided that allows the functions of hardware objects to be specified in a language such as VHDL. In this case the code produced can be treated as the first step of the implementation. In fact, an automatic translation of the MOOSE class definitions from C++ to VHDL is both possible and feasible, and it is a somewhat easier task than that of translating a complete C++ program into a concurrent VHDL description. An automatic translation of this kind fits well with the requirement that the commitment to hardware and software implementations is postponed until late in the lifecycle.

We are now close to the conclusion that the Executable Model can be regarded as equivalent to an implementation, hence we need to consider how the total effort in creating the model compares to that for creating the implementation by other means. From the software point of view we should note that the model contains much less actual software code than a stand-alone implementation, since only the basic functions have been coded and the overall structure in which they fit is synthesised from the diagrammatic model. Although other approaches to software development might use similar generator techniques, there is little margin for the approach to be improved upon. Also the uncommitted nature of MOOSE designs increases the potential for re-use and, in practice, there is a good possibility that a significant proportion of a model will be able to re-use 'library' components from earlier designs, which further reduces the work required to produce the model. We believe that not many companies will plan products in totally uncharted territory, and so they have the opportunity to capture experience in the form of re-useable class definitions.

Of course the above discussion assumes that the Executable Model is complete in implementation detail. In practice, a model will fall short of an implementation by a substantial margin, most of which stems directly from the policy of making the Executable Model suited to the codesign process, which requires the postponement of all implementation decisions except those needed to decompose the system into functional components and produce behavioural specifications for them. This gives the Transformational Codesign process the freedom to explore the full design space, as a result of which decisions will be taken that lead to the addition of new objects, which goes beyond its more obvious role of deciding in what technology existing objects are to be implemented. However, the recommendation to users of MOOSE is that the Executable Model is made logically complete and that its behaviour is validated before it is passed to the Transformational Codesign process. The effort will not be wasted and there is still some distance to go before the job is done.

6.4 THE DYNAMICS OF AN EXECUTABLE MODEL

The dynamic behaviour of the MOOSE hierarchical model is equivalent to the behaviour of the 'flat' network of primitive objects introduced above, and the most complex aspect of this behaviour derives from the concurrency of these primitive objects. By definition, all the objects in the network can operate concurrently and each may have internal concurrency arising from the simultaneous operation of its constituent functions.

Active functions require no stimulus and they operate continuously, even though they may be 'busy waiting' for changes in information flows or data stores. Therefore the base level of concurrency in a model is determined by the number of active functions it contains. The peak level of concurrency that can be reached corresponds to the simultaneous occurrence of all the possible events, since event functions are defined to execute whenever their associated events occur. In the unlikely situation of multiple simultaneous occurrences of the same event, they are assumed to be queued, which results in repeated execution of the associated event functions. Moreover, state machine functions, which are allowed to receive several events, will be entered once for each event that occurs.

For the purpose of analysing the dynamics in more detail, it is useful mentally to carry the flattening process one stage further by substituting for each primitive object the functions shown on its CID, to produce a flat network of interconnected functions. There is a possible recursive element in the substitution in that, if a CID contains class symbols, the functions on their CIDs have also to be substituted, and so on. A hypothetical

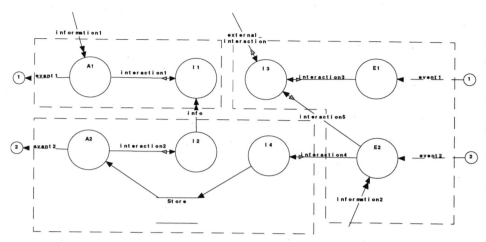

FIG. 6.6 A flattened network of functions (broken lines showing the encapsulating objects).

example of such a network is shown in Fig. 6.6, which includes active functions ('**A**'s), interaction functions ('**I**'s), and event functions ('**E**'s).

According to the MOOSE semantics, a thread of execution is permanently associated with each **A** function, although it may pass transiently to any **I** functions associated with the interactions that the **A** functions use. The **E** functions will receive a thread of execution whenever their associated events occur, which ceases when the function has been executed. As with **A** functions, the thread of **E** functions may transiently pass to any **I** functions whose interaction is used. It should be noted that event signals can enter a system model from its external interface or they can be generated internally by any function shown as a producer of the event. When an event is produced internally, an immediate increment in the number of concurrent threads is implied.

Within the individual functions, normal sequential programming conventions apply and the statements are executed in sequence so that the functions have finite, but unpredictable, execution times. They are necessarily unpredictable because, at this stage, the technology to implement the functions has not been decided. This raises questions concerning the effects of dependencies between concurrent threads, which arise in several ways. First, functions executing in parallel may have mutual producer–consumer relationships based on information flows, as shown between I2 and I1 in Fig. 6.6. Second, a 'hidden' dependence can exist between functions of the same primitive object if they transfer information via the state information of their containing object, as I4 and A2 appear to do in Fig. 6.6. Finally, any function running on one thread of execution may initiate another thread by producing an event signal, which introduces a timing dependence. For example, an **event1** is produced by **A1** which is to be serviced by **E1**.

(a)

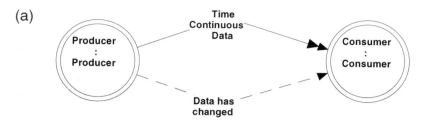

(b)

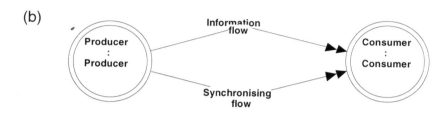

(c)

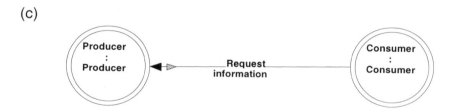

FIG. 6.7 The synchronisation of functions that interact through information flows.

In order to achieve predictable behaviour in the face of unpredictable timing, a general recommendation to users of MOOSE is that they adopt the safe policy of using explicit synchronisation between all objects and functions that *interact* through information flows, if (and only if) synchronisation is necessary to guarantee correct behaviour. This synchronisation may be achieved in several ways.

One method is for the function driving a time-continuous information flow to signal that the information flow has changed through an associated event. At the receiving object, the information flow and the event can be routed to the same (event) function, which will be executed when the event is generated (see Fig. 6.7a), which means that the earliest time of consumption relative to the change of the information can be controlled. However, this method of synchronisation may be considered inappropriate due to the increased coupling it implies between producer and consumers. For example, the object producing the change on the information flow is made aware that there are difficulties elsewhere, and in producing an event to 'strobe' the information flow, it is assuming that there will only be one consumer.

An alternative method, which eases the coupling in some respects, is to pair an information flow with a 'synchronising' information flow, so that a consumer can 'busy wait' on the value of a synchronising flow. This is in effect a 'data ready' signal that is a mechanism common in hardware. In both cases it might be necessary to have a mechanism to signal back to the producer that the information has been processed, as shown in Fig. 6.7b. Either an event or information flow might be used for this purpose, but again there is the issue of increased coupling.

The use of information flows is a common technique in hardware to provide synchronisation, such as the marking of significant changes (for example by 'data ready' signals) or interlocking operations (for example by 'handshaking' protocols). It should be used sparingly in an Executable Model because of its obvious hardware bias, although similar styles of synchronisation might be needed between the functions of a single object that uses data stores rather than information flows. In general, a more explicit synchronisation is provided by interactions, and this should be exploited as far as possible to avoid unnecessary coupling at the expense of replacing data flows by interactions. Interactions might even be developed that explicitly provide higher-level specialised synchronisation and interlocking facilities, and it will be seen that these are needed later on in the development, when the operation of some parts of a model is assigned to software that contains critical paths that must not be interfered with by other concurrent software paths.

The synchronisation obtained through interactions is that the calling function is suspended until the called function returns a result. Multiple 'simultaneous' uses of an interaction also need to be considered, and the semantics of the modelling notation specify that these interaction requests will be queued and executed in arrival order. Of course the notion of simultaneity in concurrent threads is not well defined, because of the uncertainty about the execution times of functions. Hence reliance should not be placed on assumptions about the arrival order of interactions from independent sources. In general, if sequence is important, it should be enforced by more explicit means so that the implementation is made safe regardless of arrival order. In fact, the main purpose in preventing several simultaneous executions of a function is not to aid synchronisation but to avoid the need for the designer to make function definitions safe in the context of multiple executions, and for the same reason a similar constraint is imposed at the object level to avoid the need to protect the integrity of shared state information. Thus all the interactions entering a primitive object are serviced one at a time in arrival order.

Designers of models need to apply extra care with internal function calls within an object, in case multiple simultaneous executions of these functions occur (see, for example, the **E2** to **I3** call in Fig. 6.6, which could coincide with an external request to service interaction **I3**). The internal calls are not subject to the automatic blocking applied to external entries to interaction functions, and simultaneous executions could result that might corrupt the private data of the function. Hence some explicit synchronisation needs to be implemented; for example, a function might busy wait on a flag, which is set and reset by another function. Whatever mechanism is chosen for this purpose, it will need to be reviewed later when the choices are made concerning actual concurrency in the implementation (see Chapter 9).

In summary, the concurrency semantics described above have resulted from the fact that the Executable Model represents the logical behaviour of a system before its components have been committed to software or hardware implementations, and before even the number of processors has been decided. It is required that a model performs safely even though the speed of its operations, for example, the time taken from receiving a stimulus to using or producing an information flow, is not known In an actual implementation, the operation of the functions in hardware will remain concurrent except where they explicitly interlock, whereas in the case of software objects, the actual concurrency will be limited to the number of processors used and there may be pseudo concurrency generated by interrupts, task changing and the scheduling of event processing. For example, active functions might have to run in sequence and they might be suspended in order to execute event functions.

For these reasons a compromise solution has been adopted for the dynamic semantics of a MOOSE object, in which its active functions and event functions are assumed to be fully concurrent but the number of additional concurrent threads due to entries via interactions is limited to one. The underlying assumption behind this policy of biasing towards concurrency is that it is easier to remove unnecessary concurrency when commitments are made to an implementation than it is to introduce new concurrency.

6.5 SIMULATING THE EXECUTION OF A MOOSE MODEL

As already stated, the operation of a MOOSE model is simulated by a C++ program synthesised from the information contained in the Executable Model, a significant proportion of the code being taken directly from the textual definitions linked to the graphical model. Other (simulation) objects are added, which interpret the semantics of the graphical part of the model, schedule the operation of the model, and instantiate the required objects and provide the necessary linkages between them. Figure 6.8 shows the four principle objects of the synthesised simulation program in MOOSE notation.

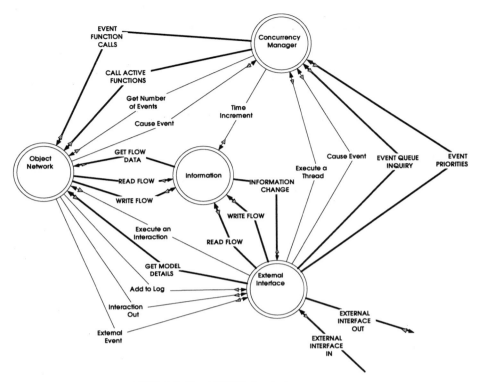

FIG. 6.8 The MOOSE simulation system.

Only one of these objects, the **Object Network**, is model-specific; it contains all the primitive objects from the Executable Model, together with some synthesised objects that provide a model-independent interface to the three simulation objects, **Information**, **Concurrency Manager** and **External Interface**. The model-independent interface consists of interactions which, through the use of pointers to actual parameters, provide the means to call every active function, event function, parameterised event function and externally accessible interaction of the primitive objects contained within the model. Other 'inquiry' interactions are provided so that model-dependent detail can be obtained; for example, the number of active functions in the model, and the number of event functions and the events that they service. It can also be seen in Fig. 6.8 that events generated by the model, external interactions that it calls, and all read and write accesses to information flows in the model, appear as interactions to be serviced by the simulation objects.

6.5.1 Internal Operation of the Executable Model

The only aspect of the internal operation of the Executable Model that needs to concern users is the way in which it simulates concurrency. Obviously a sequential program simulating the execution of a MOOSE model cannot reproduce exactly the behaviour defined by the model, to the extent that snapshots taken at a moment in time would show identical activities at identical stages of execution. An interpretative approach in which the processor time is allocated in very small time steps to each object in turn would be better, but even this would not reproduce fully the true concurrent behaviour.

Nevertheless, if the simulation is to be useful, there must be a close correspondence between the observable results it produces and those of a true implementation of the model. To understand the full significance of this problem it is necessary to have a clear view of the operation of the **Concurrency Manager** object that simulates the dynamics of the Executable Model. We can then consider the way in which this simulation relates to the specified behaviour of the model, and the possible impact of this on the observable results of running the simulation.

A variety of policies could be adopted by the **Concurrency Manager** that would produce different trade-offs with respect to complexity, run-time, and the closeness of the simulation to reality. Alternative policies are discussed in Chapter 9, but it is convenient in this chapter to fix on a simple one in order to establish the basic principles of using an Executable Model. Thus the **Concurrency Manager** object is assumed to control the execution of the **Object Network** by driving it through an indefinite sequence of *execution cycles*.

Each execution cycle starts with the user of the model being invited to supply new input values, as if from connections from the system's environment. In general, these may consist of changes to information flows, triggers for events, and requests to execute interactions. The information flow changes are relayed to the **Information** object so that subsequent accesses will receive the new values, and the triggered events are queued in the **Concurrency Manager**. Interactions are processed immediately by the **External Interface**, which uses the generic interaction (**Execute an interaction**) with the **Object Network** as shown in Fig. 6.8, and the result is returned to the user. Thus external interactions entering a model are serviced between execution cycles.

When all the required input changes have been entered and any external interactions have been processed, the new execution cycle starts. First, the complete set of active functions is executed (in an arbitrary but consistent sequence) and any events arising are added to event queues in the **Concurrency Manager**. Next, the events, both externally entered and internally produced, are serviced one at a time, by the **Concurrency Manager** invoking the associated event functions through the interaction **EVENT FUNCTION CALLS** shown in Fig. 6.8. The order in which the queued events are serviced is governed by a predetermined priority, which can be controlled by the user. If new events are signalled as a result of processing other events, they are added to the queues and become immediate candidates for execution subject to the priority rules. The execution cycle completes when the event queues are empty.

For both active and event functions, any interaction functions that they call within the **Object Network** are run to completion on the thread of execution of the calling function, and this includes outward going external interactions which, in the simulation, become interactions with the user. All outputs from the model are updated each time the **Concurrency Manager** regains control; that is, on completion of each active function or event function that it calls. For reasons discussed below, the relative timings of outputs during an execution cycle may or may not be significant.

The objective of the design policy for the **Concurrency Manager** is to allow full propagation of the effects of input changes given at the start of an execution cycle to take place during the cycle, but some difficulties arise concerning propagation of the effects of information flow and data changes made during the cycle. For example, if an active function executing late in the cycle produces information or data changes that are intended to be acted upon by other active functions that were executed early in the cycle, the effect of the information change will not fully propagate through the model until the next execution cycle. Two solutions to the problem can be offered: either the modeller must be aware of the problem and allow sufficient execution cycles to pass between input changes, or the **Concurrency Manager** must be made intelligent enough to detect when full propagation has not occurred, and repeatedly run the active functions until it has. The latter solution is not as difficult to implement as it might appear; for example, all information flow changes are already known to the simulation objects, and knowledge regarding which functions consume the flows could easily be made available. However, this has to be weighed against the extra time the simulation would take.

The only control that a user has over timing within the execution is by defining priorities for events, and hence the order in which event functions are run, and by determining the timing of inputs relative to each other and relative to the sequence of execution cycles. The actual input changes may be made interactively, or specified in a script so that long scenarios can be prepared 'off line' and used repeatedly. Output changes produced by the executing model are returned to the user interactively on completion of each active and event function, or they can be placed in a file.

The sequence in which outputs appear may only be relevant when they come from interlocked actions. Careful visual inspection of the model is needed to discover where the relative timing can be trusted. In cases where the modeller knows that the sequence is relevant, an option to provide new input at the point in the execution cycle where the output is generated can be exercised.

In conclusion, the critical questions for the user of the Executable Model are: 'How closely does the simulation follow the semantics of the Executable Model?'; 'What aspects of the model can its execution validate?'; and 'How does time in the execution map on to real time?'.

Clearly, the answers to the first two questions depend upon the degree of dependence that exists between concurrent activities. Probably the most serious perturbation of concurrency effects are as a result of the sequential execution of the active functions, which can have obvious side effects if there is information passing between them in information flows. Another possible effect of the concurrency simulation results from the sequential execution of actions stimulated by events, whereas the semantics of the modelling language expect concurrent execution. However, models that are sensitive to the order of response to virtually simultaneous events may be unsafe designs in practice, and the user must examine this issue. Since the user is allowed to know and to control the relative priorities of events, experiments can be designed to explore the significance of ordering the event processing.

The third critical question raises a broader issue concerning how real time is to be dealt with in a model, and, with the operation of the model established, we are now in a position to examine this broader issue in the next section.

6.5.2 Issues Concerning the Modelling of Time

Some objects in a system may be required to have time-dependent behaviour. For example, an object may be expected to provide a source of 'time-of-day' information, or it may be required to measure real-time intervals to meet a timing specification, such as taking samples on a periodic basis or producing output to conform to an asynchronous communication standard. The eventual implementation of these systems usually involves a device providing a periodic time signal, some form of delay, or a more sophisticated component such as a clock 'chip', which has registers that provide a symbolic time-of-day and possibly periodic interrupts. The problem is how to marry the modelling of this kind of time-dependent behaviour to the data- and event-driven semantics of the notation described thus far. A different timing issue, concerning how long it takes a function to complete and how long it takes signals to propagate (which has to be dealt with when a behavioural model is committed to implementation technology) is discussed in Chapter 9.

In a sense, the **Concurrency Manager** object has an approximate and non-linear experience of the passing of time because it has an execution cycle that passes the thread of control of the simulation program into the model, and receives it back some time later after some outstanding actions have been completed. However, no linear model of time could be inferred from this experience, nor could it be calibrated with simulated real time. At best a count can be maintained of the number of non-uniform time intervals (execution cycles) that have passed since the simulation began.

In an uncommitted model where, by policy, there is no knowledge of the time that operations take, this is the closest approximation to real time that can be provided. Such a counter can support directly the coarse modelling of the concept of 'before' and 'after', and allow an approximation to periodic time signals to be produced as events. From the periodic events a real-time object can be provided that delivers time-of-day information. However, this clock has a rather irregular tick that will be improved upon in an actual implementation by using a crystal oscillator.

This counter is made available to any function in the model in the form of a 'built-in' time-continuous information flow 'Simulated Time', and it is, in fact, used in the implementation of the VCR model's **Clock** object (Fig. 5.4). The same technique allows a crude validation of objects that have to be time-conscious, but obviously the simulation of objects that require precise time delays cannot be fully validated by this means.

6.6 USING AN EXECUTABLE MODEL

Ideally, the validation of a Behavioural Model should aim to show that the correct external behaviour is produced under all working conditions. However, the validation of even a simple model of a computer system must recognise that it is not realistic to explore the full state space of the model, either by visual inspection or by using a dynamic simulation. This problem is not a result of using models; rather, it is an intrinsic feature of computer systems arising from the fact that they almost invariably operate in a very large state space. Conventional implementations suffer the effects of this same problem to a chronic and costly extent.

Since our purpose in this chapter is to explore the use of the Executable Model to 'test' the design, a pragmatic approach will be taken in which a preparatory analysis based on visual inspection is carried out in order to identify a selection of scenarios to be 'tested'. The aim will be to provide sufficient confidence from experimentation with the model to justify continuation of the development based upon it. Further testing later on will be inescapable but, by this early testing, error density will be significantly reduced in the implementation delivered for pre-release testing.

There exists the potential for improvements to model testing, validation and verification, in general through the use of automated static analyses, which might detect some actual faults in the model but will also assist in identifying possible 'accident blackspots' that require experimental investigation. Here, we propose to start the validation of a model with a careful visual inspection, which will often detect errors but whose primary purpose is to identify scenarios to be simulated to test for correct operation of the Executable Model. Later, we will need to take account of the possibility of the simulation producing misleading results, principally due to the difficulty of simulating the concurrency in a model by software that is largely sequential. Also, we need to keep in mind that, when a model is transformed by codesign into an implementation, the concurrency realised will normally be less than the peak level that can apply in the model, and is often less than its base level. Thus more validation will be needed after implementation commitments have been made to ensure that timing deadlines can be met and that the system cannot 'deadlock' or 'livelock'. With these provisos, the analysis carried out in preparation for validation experiments should provide scenarios that lead to a sufficient and appropriate set of experiments. These can be placed in the following five categories.

1. Experiments to validate independent stimulus-response paths through the model.
2. Experiments to investigate context sensitive responses.
3. Experiments to investigate the effect of corrupt or erroneous inputs.
4. Experiments to explore potentially anomalous timing dependent behaviour.
5. Experiments to test the correct initialisation of a model.

6.6.1 Testing for Stimulus–Response Behaviour

The scenarios that will test for correct responses to stimuli can usually be identified by inspection of the External View diagram. For instance, examination of Fig. 5.2 reveals that there are only two sources of stimuli for the VCR model, which originate from the **User** and the **Tape Drive**.

The observable responses to user stimuli fall into three categories. The simplest direct responses will be changes in status information as a result of simple switching stimuli. For example, the operation of **System On** and **System Standby** should affect the **On/Standby** information flow. Other direct responses should be evident in the form of **TAPE CONTROLS** events sent to the **Tape Drive** as a result of user requests for play and record functions. Also associated with the processing of these requests, there will be feedback to the **User** through the appropriate information flows. Stimuli in the form of **OTHER CONTROLS** entering the model, which require the user to provide appropriate parameters, should generally produce immediately detectable responses in the form of changes to the information flows returning to the user. There may also be time-delayed responses in the case of automatic recording requests. The stimuli from the tape drive consist of the events **Tape In** and **Tape End**, and these should produce direct responses affecting the **TAPE STATUS** information flow.

Table 6.1 indicates the kind of simple stimulus–response scenarios that might be checked by an initial set of experiments using the Executable Model.

TABLE 6.1 Isolated stimuli scenarios for the VCR model.

Response to be Tested	Stimulus to Apply	Observed Result
On/Standby Indicator	System On	On/Standby On UHF Out from tape
On/Standby Indicator	System Standby	On/Standby Off UHF Out Nil
Time Setting	Set Time (new value)	time <- new value
Tape Playing	Play	Playing On Slow Forward Set Play UHF Out from tape

6.6.2 Testing Context Sensitive Responses

In practice, an analysis of the system model below the External View is needed in order to discover responses that might be sensitive to context. This analysis usually works well if it is centred on the first level of decomposition below the External View, but lower-level diagrams and dictionary entries will need to be consulted for detail. Context sensitive behaviour for the VCR can be identified in the **LOCK** object. For example, if the system is in standby mode, it will not respond to any input stimuli except the event **System On**. When this event occurs, other events in the bundles **PLAY CONTROLS**, **RECORD CONTROLS** and **OTHER CONTROLS** will be taken. The control of the **Tape Drive** can also be tested in a similar manner. Table 6.2 shows the system's response in a number of states to the stimulus **play** (from the bundle **PLAY CONTROLS**). Typically, the scenario would be developed to incorporate the testing of other stimuli so that the sequence of events leads the system to move from one defined state to another.

TABLE 6.2 State-dependent stimuli scenarios for the VCR model.

Response to be Tested	System State	Stimulus to Apply	Observable Results
Playing	Stopped	Play	Playing On UHF Out from tape Slow Forward Set Play
Playing	Playing	Play	No Change
Playing	Fast Forwarding	Play	Playing On UHF Out from tape Set Play
Playing	Previewing	Play	Slow Forward
Playing	Recording	Play	No Change

6.6.3 Testing Responses to Erroneous Inputs

Scenarios that present corrupt or erroneous input data to the system can be useful in assessing the safety of the system; for instance, one may evaluate the system's behaviour in the face of invalid parameters from the **User**. However, in this case, the interface will not be fully developed until the Transformational Codesign phase, and it will have to be re-evaluated and made safe again at that stage. This type of testing is also useful for systems whose behaviour depends upon processing of time-continuous information from their environments. There is no example of this in the VCR; however, the Mine Pump System provides a good illustration. In this system there are two water level sensors that should give readings consistent with one another so that, if the water level is recorded as being above the high water sensor, it should also appear to be above the low water sensor. The system's operation with a faulty sensor can be assessed, such as with the low water sensor failing and continually giving high readings. Additional checks might be included in the model to validate safe behaviour where necessary.

6.6.4 Testing for Anomalous Behaviour

The most difficult part of a validation is to assess the robustness of the model, which mainly concerns checking that no anomalous time-dependent behaviour can occur. Of course, systems that interface to the real world are presented with input changes that occur simultaneously (or at least sufficiently close together to give that appearance) or have input changes that occur when the system is processing previous input changes. The validation of safe behaviour in these circumstances is particularly important. Careful inspection of the model should at least locate potential problem areas; for example, places where stimuli leading to opposing responses converge or their paths interact are possible 'accident black spots'.

As an example of time-dependent behaviour, consider the CID for the class **Auto Recording Facility** in Fig. 6.9. The purpose of the function **start_monitor** is to compare the current **time** with the recording start time held in the store **programme**. When these times match, the automatic timed recording is started, provided that the VCR has been set in the appropriate mode, indicated here by the store **active**, which can be toggled by the functions **activate_record** and **deactivate_record**. Consider the case of the auto-recording facility being enabled at the same instance that the time changes and reaches the starting time of the programme. Depending on the relative order of arrival, either the auto-recording programme is started normally or it is ignored (since the mode has not yet been set).

The variations are, of course, due to the unknown response times of

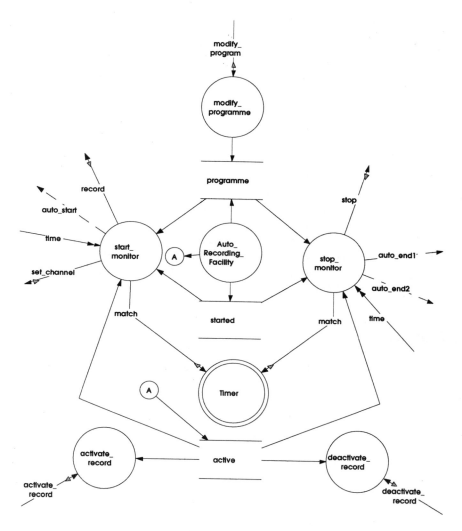

FIG. 6.9 The CID for the auto-recording facility.

the objects concerned with respect to updating internal state, signalling detected events and responding to interactions, all of which can be happening concurrently. We need first to ask the question, 'Does it matter whether the programme is started or not?'. This is really a behavioural issue, which should be resolved by knowing the intended operation of the auto-recording facility. If it is intended that the programme is started only when there is an exact match between the start and current times, the system behaves unpredictably on the simultaneous operation. Once the time match has been missed, without further action by the user the system will start recording, perfectly legitimately, on the corresponding day a week after the intended start time, because the specification stated that this VCR does not have a date, but only a day, indication. However, if it is intended that an auto-recording will start whenever the current time is *later* than the start time, the variation in arrival is unimportant, and this is indeed the way the VCR model presented here is designed.

6.6.5 Testing for Correct Initialisation

The Executable Model is useful for one further type of model evaluation, namely initialisation. MOOSE models may have constructor functions in any of their objects, and these are executed first, before a simulation run is started. In an implementation, they become the 'power up' action of a system. In other forms of behavioural models, the focus is entirely on the steady state operation of the system, and they fail to consider how the system 'powers up' and reaches a safe initial operating state.

Correct initialisation behaviour can be investigated very effectively in a MOOSE model as a natural extension of the mechanism for exploring operating behaviour.

6.7 SUMMARY

Validation of models is important if they are to evolve into implementations. This chapter has described an approach to developing and using an Executable Model that is unique to the MOOSE paradigm. The authors believe that it is effective in reducing errors and cost-effective in use; however, they would not wish to claim that it is Brooks's 'silver bullet' (Brooks 1987, Harel 1992).

The MOOSEbench tool is made available on the World Wide Web to provide readers who have access to suitable equipment with the opportunity to experiment for themselves with Executable Models. The internal operation of such a tool has been described in this chapter because the authors believe that some user knowledge is required of how the Executable Model is simulated in order to make effective use of the model. Also, the tactics for using simulations have to be guided by a good knowledge of the inner working of a model. For example, the **External Interface** object in Fig. 6.8 in particular plays a complex role that must be understood. The most obvious part of this role is to provide the line of communication with the user through which outputs from the model are made visible and new inputs are captured. For the simulation policy outlined above and included in the tool, the interface has to allow a user to supply input at two key times: at the start of an execution cycle, and just after an output has occurred. The presentation and prompting of information can be configured by the interface; for example: to prompt for user input only after a specified number of execution cycles; to obtain user input from a file; or to await input after each output is produced.

As well as collecting and displaying information to and from the executing model, the interface allows the user to have some control over the execution, and it also provides debugging facilities. Examples of this control are: the user can stop the execution on certain conditions (by setting breakpoints); and the user can change the priority of the events in a model.

7 Designing to Meet Constraints

In earlier discussions we emphasised that the developers of computer systems products must deliver systems that meet both the Functional Requirements and the Design Constraints imposed on the system, and examples of systems that failed to do so were presented in Chapter 1. So far the discussion has centred upon the development of models that focus on the Functional Requirements of a system. We will now turn our attention to the way in which Design Constraints are addressed in the MOOSE paradigm. In particular, we are concerned with identifying how and where they affect the decisions that are taken which determine how well a system's implementation meets the imposed Design Constraints.

As discussed in Chapter 5, a MOOSE development begins by categorising the system requirements into Functional Requirements (FRs) and Design Constraints, the latter including Non-functional Requirements (NFRs), Design Objectives (DOs) and Design Decisions (DDs), and the phases of the MOOSE paradigm, in which each category of requirement exerts influence on the system model, are shown in Fig. 7.1. This indicates that both the Functional Requirements and Design Decisions are used to shape the development of the product architecture and, in particular, the Behavioural Model. The role of Functional Requirements in this respect has been explored in Chapter 5, but the need to account for Design Decisions at this early stage requires a brief comment. As indicated in Chapter 5, externally imposed Design Decisions are unavoidable in the development of many Computer Systems products, and so they should be taken into account as early as possible in the construction of a system model to ensure that the model, and ultimately the implementation, conform to the decisions.

Once the Behavioural Model is complete, it is reviewed against the Functional Requirements and Design Constraints. At this stage, the review will typically take the form of a 'walkthrough' of the model, where the system's responses to expected patterns of input are investigated by manual means, and unsatisfactory behaviour will lead to modifications to the model. As part of this process the Design Constraints will be considered, and this may again lead to changes in the model structure. Modifications which may be made in the light of the Design Constraints

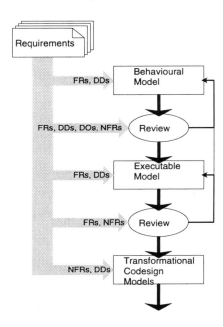

FIG. 7.1 The influence of design constraints on the MOOSE development process.

include restructuring to reduce the length of time-critical paths through the model or the number of possible failure modes and safety critical components. Such modifications are discussed later in this chapter.

Once the Behavioural Model is deemed to be satisfactory, it is developed into the Executable Model, as discussed in Chapter 6. Again, externally imposed Design Decisions requiring, for example, the use of a specific algorithm, may affect the development of the model. Upon completion, the Executable Model is reviewed against the Functional Requirements (see Section 6.6) and against the Non-functional Requirements, since some information about issues such as timing and code size may be derived from the Executable Model. A process of model refinement may be undertaken if the review reveals significant flaws.

The next stage of the MOOSE paradigm is Transformational Codesign, where the uncommitted Executable Model is transformed into an implementation. This process requires many decisions to be taken, including:

- Selecting appropriate interfacing mechanisms.
- Choosing the implementation technology (hardware or software) for all the primitive objects.
- Choosing the number and type of processors to be used in the implementation.
- Assigning objects to particular processors.

- Selecting the technology for interconnecting the processors and other hardware objects.
- Selecting the most appropriate form of run-time support for the software objects.

These, and other related issues, are considered in Chapter 8, where the process of Transformational Codesign is discussed in detail. However, it is clear that all of these decisions will be strongly influenced by the system's Non-functional Requirements, and at each step in the commitment process the Non-functional Requirements constrain the designer, filtering out possible system commitments which do not satisfy the Design Constraints.

In this chapter we will briefly review methods for reasoning about Non-functional Requirements during system development, and then consider how certain Non-functional Requirements may be used to modify MOOSE Behavioural and Executable Models. There is also a role to be played by the Non-functional Requirements in the later phases of MOOSE, specifically Transformational Codesign as described in Chapter 8. The methods for reasoning about Non-Functional Requirements that appear below can also be applied in this phase of the paradigm.

7.1 CONSTRAINTS ON THE DESIGN PROCESS

A number of constraints on the design process were identified in Chapter 5, namely Non-functional Requirements, Design Objectives and Design Decisions. It is worthwhile to note that this is not the only possible classification of Design Constraints. For example, Sommerville (Sommerville 1996) classifies constraints as Product Requirements, Organisational Requirements and External Requirements. Product Requirements include what we have termed Non-functional Requirements and Design Objectives. These originate directly from the customer and characterise the desirable attributes of the product. Organisational Requirements basically constrain the developer to adhere to particular standards relating to, for example, methods, languages and tools, and are usually imposed by the developers' organisation or by the customer. Finally, External Requirements represent a catch-all category which includes issues such as those concerned with interoperability requirements, budget, time scales and legislation. In terms of the classification used in this book, and depending upon the exact nature of the requirement, External Requirements would be categorised as Non-functional Requirements, Design Objectives or Design Decisions.

Although neither of the classification schemes identified is innately superior to the other, we feel that distinguishing constraints which must be met and are quantifiable, from those which may be traded-off and are unquantifiable, is advantageous.

7.2 EVALUATING NON-FUNCTIONAL REQUIREMENTS

Approaches which provide a framework for the consideration of Non-functional Requirements within the development lifecycle are rare, although many approaches have focused on the consideration of a single kind of Non-functional Requirement during development: primarily timing and performance, and to a lesser extent reliability. Representative approaches will briefly be considered below, first for software systems and then for hardware. MOOSE, however, aims to be able to allow all relevant Design Constraints to be accounted for during development, and so in Section 7.4 attention will be focused on approaches which provide a general framework for the analysis of Non-functional Requirements and the assessment of their impact on a system design.

Ensuring that a system design meets its Non-functional Requirements is currently one of the most problematic areas of system development. One reason for this is that the set of Non-functional Requirements which are important to a particular system's development are clearly application-specific. For example, in space systems or hand-held products, the size, weight and power consumption are all clearly critical, whereas for many other systems they are at most of secondary significance. This factor, along with the difference in nature of the various Non-functional Requirements, has meant that specific Non-functional Requirements are accounted for in development by specialised approaches and so there are few frameworks for handling all Non-functional Requirements relevant to a system.

For software systems, many approaches have been developed which are capable of dealing with single Non-functional Requirements. Timing requirements may be evaluated by approaches such as timed Petri Nets (Ghezzi et al. 1989), timed CSP (Davies et al. 1992) and TLOTOS (Leduc 1991), and later in development, when a tasking structure for the application has been established, tasks may have deadlines associated with them and a schedulability analysis may be carried out in the context of the kernel which is to be used by the system (Audsley et al. 1993).

The performance of a system may be analysed by analytical methods or through the use of discrete event simulations (Jain 1991), although the latter is probably more popular due to the availability of simulation tools such as SES\workbench.

Developers may seek to ensure the desired levels of reliability through fault tolerance, using design, implementation and testing techniques such as N-Modular Redundancy (Avizienis 1985), recovery blocks (Randell 1975) and other defensive programming strategies. Reliability growth models (Abdel-Ghaly et al. 1986) may also be used on an implementation to predict when the required level of reliability is likely to be attained through debugging. Safety requirements are typically dealt with by fault tree analysis (Leveson 1986), although formal methods are also increasingly used for the high integrity components of safety critical systems (Barroca and McDermid 1992).

The Non-functional Requirement affecting the largest number of diverse projects is perhaps cost and, accordingly, a great many approaches to cost estimation have been proposed. These range from judgements made by experienced personnel through to algorithmic methods. Algorithmic cost modelling is perhaps the most methodical (although not necessarily the most accurate) approach and assumes that cost is an empirically determined function of a number of project metrics related to size, functionality, complexity and the development process. Size, in terms of predicted lines of source code, is very frequently used in algorithmic cost modelling, although it is commonly accepted that this is a highly imprecise metric (Sommerville 1996). Code size is usually estimated at the start of a project with reference to previous projects, the historical data being used to reason by analogy about the expected size of the code for a new application. In addition to its use in cost modelling, code size estimates may also be used for predicting memory requirements for both code and data (Lawrence and Mauch 1988).

The approaches outlined above fall into one of three categories:

- System representation notations
  These include timed Petri Nets, timed CSP, TLOTOS, discrete event simulations, and analytic performance models.
- Design/implementation/testing strategies
  These include N-Modular Redundancy, recovery blocks, and reliability growth models
- Design analysis
  Schedulability analysis and fault tree analysis.

The system representation notations may all, in principle, be used as a basis for complete computer system development, since their advocates suggest that they are applicable to both hardware and software. However, no well defined and widely used development paradigms appear to have been devised based upon these notations. Moreover, timed CSP and TLOTOS are textual (formal) notations and, although timed Petri Nets have a graphical representation, they suffer from difficulties analogous to the state explosion problem encountered when trying to represent systems of even modest complexity by finite state machines (Harel 1987).

The design/implementation/testing strategies outlined above may be applied in any development paradigm. This is also true of design analysis, although in the case of schedulability analysis, the semantics of the paradigm's supporting notations may need to be restricted in order to ensure that the analysis is tractable (Burns and Wellings 1994).

However, all the approaches outlined above have virtues, and so it is desirable to consider briefly how they may be integrated into a system development paradigm. One suitable view would be to regard all of the above approaches as techniques to be applied within a method step that is concerned with evaluating whether a system design meets the imposed Non-functional Requirements. In the case of the system representation notations, some form of mapping from the primary development notation (for example MOOSE) into a special purpose notation would be necessary, and similar work with other methods has been undertaken, notably with Petri Nets being used with HOOD models (Robinson 1992), and timed Petri Nets being derived from models in the MOON notation (Hull and O'Donoghue 1994). Indeed, some work has already been undertaken in mapping MOOSE models with TLOTOS. In a similar vein, SES\workbench simulations may be derived from Real-time Structured Analysis models (Burton 1991).

With respect to hardware systems, a number of Non-functional Requirements are treated using similar techniques to those described above for software. Indeed, most of these common approaches originated in the hardware sphere and have subsequently been adapted for use with software systems. This is particularly true of the treatment of performance, reliability and safety, where discrete event simulation and analytical modelling, N-Modular Redundancy and fault trees have all been used.

In the context of hardware systems developed using a modern hardware description language such as VHDL, it is common for logical timing issues to be addressed during the development of an executable model, using facilities provided by the development environment. More detailed investigation of timing is carried out during the development of a synthesisable VHDL model, and the whole model can be simulated, again using facilities commonly provided by complete VHDL toolsets.

It is usually acknowledged that 'design sizing' should be carried out at an early stage of chip development to estimate a number of parameters, including circuit size, pad count and the number of major functional elements that are available as predesigned macros for the implementation technologies under consideration (Dillinger 1988). The estimates allow the cost and feasibility of the design to be predicted. However, in practice, there appear to be few approaches for performing design sizing at the system level.

7.3 FRAMEWORKS FOR EVALUATING NON-FUNCTIONAL REQUIREMENTS

The first approach to be considered which accounts for Non-functional Requirements in the development process is due to Mylopoulos et al. (1992), with further work being reported by Chung et al. (1995). Their aim is to develop a method of justifying design decisions that affect a system's ability to meet its Non-functional Requirements. The approach is influenced by work on decision support systems and on the development environment, and is aimed particularly at information systems.

The method represents Non-functional Requirements as goals which are to be *satisficed* through the taking of appropriate design decisions during development. The term *satisfication* is used to indicate that an information system will meet the Non-functional Requirements to 'within acceptable limits', rather than absolutely. Besides goals relating to Non-functional Requirements, there are also goals linked to design decisions that influence the Non-functional Requirements, and goals that represent evidence supporting or denying the other kinds of goals. Additionally, goals may be refined into subgoals, the satisfication of which lead to the satisfication of the parent goal. Although the approach has been demonstrated with Non-functional Requirements that are stated qualitatively, and so would be categorised as Design Objectives by the classification scheme outlined in Section 5.2, there does not appear to be any reason why the method should not be used as a framework for considering quantified Non-functional Requirements.

Overall, the approach seems to provides a comprehensive framework for evaluating the impact of design decisions on an information system's ability to meet its Non-functional Requirements. However, the qualitative nature of the concept of satisfication appears to be at odds with the need, in many real-time systems, to be able to quantify the extent to which a system meets its Non-functional Requirements. Moreover, the fact that the approach is almost completely divorced from a system's architecture makes it difficult to envisage how it could be applied in the context of a paradigm like MOOSE. It also suffers from the disadvantage that, since its conceptual framework is completely independent of any development method, designers need to learn and apply two different sets of methods and notations.

A different approach, known as TARDIS (Timely And Reliable Distributed Information Systems), is proposed by Burns and Lister (1991), and Fidge and Lister (1992). This is a framework for catering for Non-functional Requirements during software development and is specifically tailored for real-time systems. Its authors claim that TARDIS can be used with any software development method and illustrate its application in the context of a highly simplified Object Oriented approach, and in software development using the formal specification language Z.

The TARDIS approach adopts the view that developers of real-time software must account for the software's execution environment (including the processor, kernel and communication mechanisms) if the system's Non-functional Requirements are to be met. They also assume that the execution environment is fixed before software development begins.

The approach recognises that the introduction of large amounts of low level detail at an early stage in development may overload the designer, and so the approach advocates a two-stage development process. In the first stage, a logical model of the software is constructed (the 'logical architecture') which is based almost exclusively on the Functional Requirements. The logical architecture represents a simple class model in the case of the Object Oriented approach used by Burns and Lister (1991), and a conventional Z specification in the work of Fidge and Lister (1992).

Once the logical architecture is complete, the physical architecture is derived from it by considering each of the appropriate Non-functional Requirements in turn, and the constraints placed upon the system by the target execution environment.

The process of developing the logical and physical architectures is represented in terms of the transformation of *obligations* into *commitments*. Obligations represent features of the system design which have not been fixed and which initially represent the client requirements, both Functional and Non-functional. The decision to utilise a particular mechanism to meet an obligation represents a commitment that designers working at a more detailed level are not free to change. During the construction of the logical architecture, the primary sources of obligations are the Functional Requirements. However, the satisfaction of obligations by commitments is subject to the constraints of the execution environment, and during the development of the physical architecture, the process of making a commitment must evaluate the possible design alternatives in the context of these constraints and the Non-Functional Requirements. This point is illustrated in Fig. 7.2, which is adapted from Burns and Lister (1991).

The TARDIS approach has been illustrated using the Mine Pump system discussed in Chapter 4, and Burns and Lister (1991) shows how Non-functional Requirements relating to dependability (safety and reliability), timeliness and distribution can be used in the development of the physical architecture. No guidance concerning how the Non-functional Requirements should influence the physical model is provided, and arguments to support particular commitment decisions are *ad hoc*. Fidge and Lister (1992) present an extremely simple embedded data acquisition system as an example, and show how Real-time Logic (RTL) can be applied within the TARDIS framework for the evaluation of timing requirements. Again, there are no guidelines about how the Non-functional Requirements should be evaluated, and the choice of RTL is arbitrary.

To summarise, TARDIS provides a framework for evaluating design commitments in the context of Non-functional Requirements. It is flexible enough to be integrated into many real-time development methods and potentially enables the impact of all Non-functional Requirements and interactions between them to be evaluated. The approach is reasonably compatible with MOOSE, although the assumption of a prescribed execution environment is one major source of discord. The need for *ad hoc* reasoning about Non-functional Requirements could be viewed as a

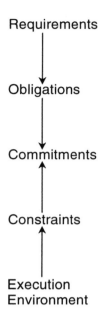

FIG. 7.2 The TARDIS model of real-time software development.

shortcoming, but there do not appear to be any other approaches in the literature that allow all relevant Non-functional Requirements to be considered during real-time software development.

More recently, Burns and Wellings (1994) have combined and extended a number of notions from TARDIS and the HOOD design method, resulting in the HRT–HOOD method (for Hard Real-time HOOD). The primary goal of the method is to produce designs for hard real-time systems that are amenable to timing and schedulability analysis. A number of new object types, beyond the active and passive objects of the conventional HOOD method, are introduced which, it is claimed, represent common hard real-time abstractions. The aim is to help the developer to produce designs that can be analysed. The method is also illustrated using the Mine Pump system and, in a separate paper by Burns *et al.* (1993), by a redesign of the control software for the attitude and orbital control of the Olympus satellite.

HRT–HOOD retains the notion of both logical and physical architectures that originated in TARDIS, although it is acknowledged that the development of both architectures may be concurrent. The logical design activity corresponds to the construction of a HOOD model as far as terminal objects that require no further decomposition. Although this process is largely driven by the system's Functional Requirements, its authors indicate that the decomposition of the logical architecture will be strongly influenced by the existence of timing requirements.

Design of the physical architecture requires the mapping of the application on to the prescribed execution environment. For hard real-time systems, the mapping must allow the resulting design to be analysable, and so it requires knowledge of the characteristics of the underlying execution environment. During the mapping, objects are allocated to processors, the scheduling of tasks on processors is considered together with the scheduling of communication between processors. Finally, dependability issues such as object replication are addressed. The method's authors claim that the approach could also be extended to consider other Non-functional Requirements such as power consumption, memory size and even weight, although they provide no indication of how this might be achieved.

While the HRT–HOOD method represents a positive contribution to the systematic development of hard real-time software, it is still founded upon the premise that hardware and software development are distinct, non-interacting activities, and so it is open to the criticisms of such approaches that were raised in Chapter 1. Moreover, although it is claimed that the method is implementation-language-independent, it is heavily biased towards the Ada-95 language, which has only recently been standardised and for which there are few commercially available, industrial strength compilers. In addition, as with many methods, HRT–HOOD suffers from the fact that accurate timing information is only available late in the development lifecycle. Thus it is quite possible that, when such information finally becomes available as each component is implemented, it will demonstrate that the original estimates were highly inaccurate, necessitating the reworking of the timing analysis at a point when there is little time left before the system is due to be delivered. The method's authors do, however, argue that the re-use of existing, well characterised objects can significantly improve the quality of early timing estimates.

7.4 NON-FUNCTIONAL REQUIREMENTS AND THE MOOSE PARADIGM

The discussion in Section 7.1 outlined, in general terms, how Design Constraints affect the development of system models. In particular, we saw that Non-functional Requirements can influence the development of the Behavioural and Executable Models, and also the Transformational Codesign. This statement needs qualification, however, since not every Non-functional Requirement affects the model at each of these stages. In particular, implementation dependent issues like hardware size and power consumption may only be considered in detail in the

Transformational Codesign phase, whereas requirements relating to timing and reliability may be examined at the end of the Behavioural and Executable Modelling stages, as well as during Transformational Codesign. However, the nature of the problems to be faced is similar; we need to look ahead to predict the consequences of decisions captured in the model with respect to the system's ability to meet its Non-functional Requirements.

In the following section, we briefly consider a number of important Non-functional Requirements which may influence the Behavioural and Executable Models. These are specifically timing, safety, reliability and software size.

7.4.1 Timing

In this section we consider how timing requirements may be used as criteria to assess MOOSE Behavioural and Executable Models. However, first we will briefly consider the nature of common timing requirements in real-time systems. These have been discussed by Gerber, Hong and Saksena (1995), who indicate that most constraints are end-to-end in nature, meaning that they specify the timing relationship between a system's inputs and outputs. Inputs will trace a path through a system's components as they are used to compute outputs and, for the overall end-to-end constraints to be met, the system components on time-critical paths are subject to intermediate timing constraints. Thus: '...a small handful of end-to-end constraints may – even in a modest system – yield a great many intermediate constraints' (Gerber, Hong and Saksena 1995).

The partitioning of end-to-end constraints between system components is difficult, since there are many ways of dividing the former among the latter and, early in the design process, insufficient information is available to developers to enable them to assess the quality (or even feasibility) of the partitioning that they have specified.

Three sub-categories of end-to-end timing requirement have been identified. Propagation Delays, or Freshness Constraints, specify the elapsed time taken from the receipt of an input to the production of the corresponding output. Correlation Constraints are specified in circumstances where two (or more) inputs are needed for the calculation of an output, and the inputs must be gathered within a certain time period. Finally, Separation Constraints are used to bound the rate of delivery of a particular output and are necessary to control jitter.

In MOOSE, timing requirements may be taken into account at various stages of the development. During the review of the Behavioural Model, system timing requirements may be used to identify time-critical paths through the model at the primitive object level. The ability of the objects on a time-critical path to meet the associated timing constraint may then be assessed. This will rely heavily upon the experience of the developers, since little detailed timing information is available at this stage. However, if the developers believe that there may be difficulties in meeting a timing requirement, it may be possible to restructure the Behavioural Model to shorten the length of the associated time-critical path.

Once the Executable Model has been developed, more detailed information is available to assess the timing characteristics of the model. In what follows, we show how timing relationships may be written for MOOSE models and indicate how some of the timing data may be acquired. This can give an early warning of timing problems and possibly lead to a restructuring of the Executable Model.

In assessing Propagation Delays, the approach requires the identification of time-critical paths through the model at the function level, and the association of timing constraints with those paths. For example, we might wish to investigate the timing requirement associated with pump shutdown in the event of dangerous levels of methane. A timing requirement T_{SD} will have been specified by the customer or its advisors. We can see, by reference to Fig. 7.3, that the **Gas** object senses the methane concentration and notifies the **Pump Driver** when there is an explosion risk.

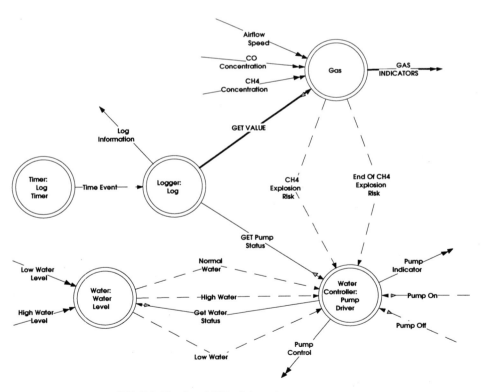

FIG. 7.3 Top-level OID of the mine pump system.

The timing requirement can therefore be partitioned into T_{GAS}, concerned with identifying when the concentration exceeds a predefined threshold, and T_{PD}, concerned with driving the pump control when there is an explosion risk. Furthermore, we can say that

$$T_{SD} = T_{GAS} + T_{SD}$$

and, as the process is iterative, the requirements can be further refined as we proceed down the OID hierarchy. Hence

$$T_{GAS} = \tau^{c}_{ch4} + \tau^{f}_{thres} + \tau^{c}_{er}$$

where

τ^{c}_{ch4}	is the time between the methane level becoming critical and the time-continuous information flow **CH4 Concentration** entering the system.
τ^{f}_{thres}	is the time to execute a function to compare the gas concentration with the threshold.
τ^{c}_{er}	is the time to propagate the event **CH4 Explosion Risk**.

$$T_{PD} = \tau^{f}_{control} + \tau^{c}_{cmd}$$

where

$\tau^{f}_{control}$	is the time to execute the pump control function.
τ^{c}_{cmd}	is the time to communicate the pump off command to the pump hardware.

Terms superscripted with an 'f' represent the time to perform the associated function, while those superscripted with a 'c' represent communications/invocation times. Although this appears to be simple enough, some care is needed in its interpretation. τ^{f} terms, relating to functions invoked by an interaction or an event, give the time taken from the instant that the function starts to execute to the instant that it completes its computation. For interactions, τ^{c} terms represent the time taken to invoke the operation, preparing it for execution, whereas, for events, τ^{c} should be interpreted as the time taken for the recipient to detect that the event has occurred, and to prepare the associated event function for execution. Although not used in this example, τ^{c} is also used to represent the time taken to return control to the caller of an interaction.

The cases of active functions and time-continuous information flows are slightly more complex, partly because these abstractions can be implemented in a variety of different ways. With respect to the function which checks a signal from outside the system and raises an event if its value exceeds a certain threshold, τ^f represents the elapsed time between the signal exceeding the threshold and the output event being generated. In cases where a function outputs a time-continuous flow, τ^f represents the latency between a change in the input time-continuous flow being reflected in the output flow.

An additional question that arises is whether a τ^c value should be associated with a time-continuous flow. The answer is 'yes' since, in a hardware implementation of a time-continuous flow, τ^c will represent signal propagation delay, or something similar. However, in the case of a software implementation, τ^c will vanish as the time taken to write the value by the sender will be subsumed in the sender's τ^f and the time taken to read the value will be part of the receiver's τ^f. It should be noted that some of the τ^c associated with interactions may disappear if, in a software implementation, their associated functions are specified to be in-line.

The above is a simple situation where one input is used to trigger a single output. More complex cases will involve multiple inputs being combined to produce an output, leading to several time-critical paths which merge at the object which combines the inputs to compute the output. Clearly, the approach described above can be extended to deal with more complex situations.

The case of several inputs being combined to produce an output leads naturally to the consideration of Correlation Constraints, where the inputs must be gathered within a specified period of one another. An example of this kind of situation is illustrated in Fig. 7.4.

Here, when object O3 receives the event E1, it calls functions O1.f1 and

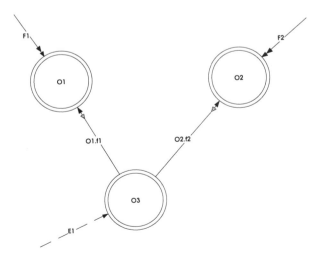

FIG. 7.4 Correlation constraints.

O2.f2 to sample flows F1 and F2. It is also specified that the two samples must be acquired within δt seconds of each other. This correlation constraint may be expressed as:

$$\delta t \geq \tau^f_{E1} + 2\tau^c_{f1} + \tau^f{}_{f1} + 2\tau^c_{f2} + \tau^f_{f2}$$

where the factor of 2 in the τ^c terms represents the call of an interaction and the return of control. This represents a worst-case condition where O3 calls for the sample from O1 at the start of event function E1, and requests the sample from O2 at the end of E1. It also assumes that F1 is sampled at the beginning of f1, and F2 as f2 completes its execution. It is necessary to use pessimistic assumptions because timing is specified at the function level, which is too coarse to capture the reality that the actual time between the acquisition of the samples will depend on the relative position of the calls to f1 and f2 from within E1, and the positions within f1 and f2 where the inputs are sampled.

The third form of timing constraints, namely Separation Constraints, relate to the rate at which a function can update an output, and therefore will take the form of constraints on the execution time of the function which delivers the output, or on the frequency with which the function can be invoked, or on both.

It is important to understand that time constraint relationships do not necessarily contain all the necessary terms at the end of the behavioural modelling stage of MOOSE. This is because additional objects that typically play an interfacing role may be introduced into the system during Transformational Codesign. Time constraint relations will need to be modified in such circumstances and Chapters 8 and 9 discuss these issues further.

Since MOOSE Executable Models may be run and the behaviour of the models observed, timing data (τ^f values) may be obtained by monitoring the execution times of the functions under a variety of conditions, and typical values obtained. They may be in terms of simulated time, as discussed in Chapter 6. Equally they may be actual measurements obtained from running the model. Although these timing values relate to the speed of the processor which runs the model, rather than the processor to be used in the final implementation, the relative timings are likely to be representative, and if the target processor is known at this stage, then the results obtained from running the model may be scaled to give estimates of execution times on the target machine. It should be noted that the timings obtained in this way relate to software implementations of the system's objects, some of which will be realised as hardware in the final system. However, such timing information, along with the time constraint relations described earlier, may be carried forward into Transformational Codesign to guide the choice of implementation technology.

As indicated earlier, the timing data and time constraint relations may also be used at the end of the Executable Modelling stage to detect if problems in meeting the timing constraints are likely to occur. If for a particular time-critical path, for example, the sum of the compute times established through model execution is close to the timing requirement, then clearly the system is unlikely to be able to meet this timing requirement, since other factors (communication times and interface objects) have been left out of the equation.

Similar considerations apply to correlation constraints. In this event there are two possible courses of action. The first is to do nothing other than indicate that, during Transformational Codesign, the objects on this time-critical path should be considered as candidate hardware objects. Alternatively, the behavioural model may be restructured, with attempts being made to reduce the number of objects/functions on the time-critical path, or to shorten the execution time of the remaining objects/functions.

The timing constraint relations and function execution time data relate to the abstract behavioural model, which in general will exhibit a high degree of concurrency. When the model is committed to an implementation, a number of additional timing issues arise, particularly for the objects that are committed to software. One key issue is pre-emption. The timing relations developed above assume that there is no pre-emption, and so functions on a time-critical path all execute to completion and pass on the thread of control to the next object on the path. However, if the committed model allows for pre-emption and perhaps multitasking, then the time taken to perform context switching and to execute pre-empting functions must be taken into account, as must the possibility that threads may be blocked awaiting access to a shared resource. Such considerations are used to supplement the basic timing model outlined above in the Transformational Codesign step termed 'Check Integrity of Threads of Execution'.

7.4.2 Safety and Reliability

Once the Behavioural Model has been completed, it may be examined from the perspectives of safety and reliability, and possibly modified in the light of the analysis. In this respect, the discussion in Burns and Lister (1991) of the development of the Mine Pump system is illuminating, and provides some guidance about how considerations of safety and reliability may be introduced into a MOOSE development at a comparatively early stage, although they must again be considered during the Transformational Codesign stage.

The component failures which may compromise the safety of the mine pump system are identified in Burns and Lister (1991), accompanied by a brief discussion of the different failure modes. In terms of the MOOSE model presented in Chapter 4, safety is threatened by the failure of the **Pump Driver** and **Gas** objects, and by the failure of the communications medium between them. As Burns and Lister point out, one way of improving the safety characteristics of the system is to reduce the number of safety-critical components. A simple way of achieving this is to require that the **Pump Driver** requests the methane level whenever the pump is to be switched on, and periodically while the pump is running. In MOOSE terms this would mean replacing the two events **CH4 Explosion Risk** and **End Of CH4 Explosion Risk** by an interaction called by the **Pump Driver**, which requests the methane level from the **Gas** object. Alternatively, the **Gas** object may make the methane level available as a time-continuous flow. However, both cases imply a polling of the methane level by the **Pump Driver** object.

The safety requirement on the **Gas** object may therefore be relaxed, to the extent that the **Gas** object can be trusted to 'fail silent'. However, as indicated in Burns and Lister (1991), it is possible that the **Gas** object, which contains the appropriate sensors, will 'fail noisy' and will provide erroneous values rather than no values at all. Although the issue of sensor failure affects safety, it is primarily a reliability issue, and so will be dealt with in the following paragraph.

The reliability requirements placed on the mine pump relate to the unnecessary loss of working shifts due to system failures, and again these issues are discussed by Burns and Lister, who quote typical reliability requirements. One particular source of failure would be the sensors, and so N-Modular Redundancy (NMR) techniques may be used in an effort to ensure a degree of fault tolerance. NMR can easily and clearly be introduced into MOOSE models through the use of the replication notation described in Chapter 4. Figure 7.5 shows part of the decomposition of the **Gas** object, in which the **CH4 Sensor** is replicated in triplicate, and the **Comparator** object implements the voting strategy. Similar structures could be used for the other system components which need to be replicated in order to meet the reliability requirements. Note that, in a real

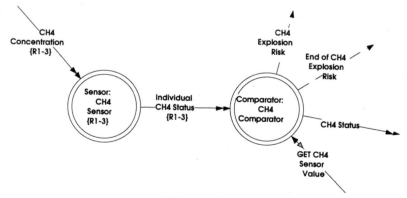

FIG. 7.5 An example of the representation of n-modular redundancy in MOOSE.

system, the comparator object would need to inform the operator in the event of a perceived sensor failure, but this is omitted in the interests of simplicity.

It should be noted that replication of the gas sensors enhances safety as well as reliability, since it is highly improbable that two sensors would fail noisily and generate the same incorrect readings.

7.4.3 Size

Upon completion of the Executable Model, an indication of the size of each of the system's objects can be obtained, in terms of code size and run-time memory requirements. However, there will be two major sources of inaccuracy in this estimate. First, the model contains software for those system components which will eventually be committed to hardware, leading to an overestimate in the size of the software. Second, no system software to support the execution of the application software will be included (either Thread Managers and hardware/software interface objects, or a real-time kernel – see Chapter 8) in the model, leading to an underestimate in software size. The first inaccuracy is difficult to compensate for at this stage of development, although, even so, it may be possible for an experienced designer to define a rough hardware/software partition on the grounds of the perceived complexity and speed of objects. It would be possible to account for the second source of inaccuracy by reference to existing system software, or to propose the use of library objects, whose sizes are known. Hence an estimate of software size can be obtained from the executable model. As with the timing data described above, such estimates may simply be carried forward and used during Transformational Codesign or, in cases where the estimates approach or exceed the maximum memory size specified in the Structured Requirements, the Executable Model may need to be reappraised.

8 Partitioning and Detailing a Computer System Design

This chapter describes the third major phase of the MOOSE paradigm, the purpose of which is to produce an implementation specification from an Executable Model by a process termed Transformational Codesign. The transformations are based on implementation decisions that are concerned with choosing the 'best' means for realising various aspects of the functional behaviour explicit in the Executable Model, in the context of the Design Constraints that are documented in the Architecture of the product. For this reason, both the Executable Model and the Architecture are inputs to the Transformational Codesign phase.

Although the Design Constraints can have a very significant effect on the actual implementation decisions taken for a product, as discussed in Chapter 7, they need not affect either the sequence in which such decisions are taken or the requirements of the notation used to mark the resultant transformations on the model. Consequently they will not be considered further in this chapter. Instead, this chapter concentrates on the nature of the transformations.

In the interest of keeping the description simple, we will consider only one class of target hardware/firmware configuration; that is, we shall assume that the required hardware platform is to be custom built, as an assembly of processors and memories combined with the hardware implemented objects from the Executable Model. We shall also assume that no operating system kernel is to be used so that the application software, which consists of the software implemented objects from the Executable Model, has to run in a 'bare machine'.

Obviously, in many practical applications the use of standard operating system software will be desirable, if not mandatory, and a less integrated hardware platform might be preferred, such as a network of standard computers supplemented by application specific PCBs. In fact the use of these less primitive platforms simplifies the implementation work, but its inclusion here would complicate the discussion of the fundamental principles of Transformational Codesign. Chapter 9, which is concerned with the pragmatics of applying MOOSE, re-visits the target platform issue and also considers a related issue, namely that of separating the software into application software and application-specific kernels.

Clearly there are many benefits to be gained from producing an implementation by applying transformations to a validated Executable (Behavioural) Model, rather than by creating the implementation detail in a more independent way and involving the risk of introducing incorrect or undesirable behaviour into the product. Thus the goal of the Transformational Codesign process is to produce an implementation specification in the form of models developed directly from the Executable Model. When complete, these models form the input into an automatic process that synthesises implementation source code. We shall assume that the hardware implementation is to be from a VHDL specification* and that the software is to be in the form of separate C++ programs for each processor contained in the hardware platform.

8.1 THE METHOD OF TRANSFORMATIONAL CODESIGN

The method for producing the implementation specification is termed Transformational Codesign because it is based on a prescribed series of transformations, which apply decisions concerning the implementation of the hardware and software objects in the Executable Model. These transformations may add some new implementation-dependent objects to the model, modify some inter-object connections, and mark objects in the model to show whether they are to be implemented as hardware (*hard* objects), software (*soft* objects), or software that will be very hardware- or firmware-dependent (*firm* objects). Examples of each are given later. At the end of the transformations a complete commitment will have been made to the method of implementation of every object in the Executable Model. The result is called a *Committed Model* and it forms one part of the implementation specification. In general, experiments with the Executable Model might be needed to confirm the effectiveness of the commitments to hardware and software, and so the model remains executable throughout its transformation.

After the Committed Model has been produced, another model called the *Platform Model* is automatically created from it. The main purpose of this is to facilitate the development of further hardware detail and to specify how the hard objects are to be connected physically to form the support platform for the application specific software. Initially it will contain all the hard and firm objects identified on the Committed Model and some new firm (interface) objects to encapsulate the implementation specific detail of accessing the hard objects. The final stage of Transformational Codesign transforms the initial Platform Model into a complete physical specification of the hardware (and the 'firm' software) that is to be built as the plat-

---

* Other forms of hardware specification are considered in Chapter 9.

form on which the application specific software will run, thereby completing the implementation specification of the total system.

In the case of both the Committed Model and the Platform Model, the transformations should be either an additive or strictly localised change. If any transformation leads to a significant disturbance of the validated detail of the model, the structure of the Executable Model should be revised and re-validated.

In the next and final phase of the MOOSE paradigm, the hardware source code is synthesised from the Platform Model. The software source code is largely synthesised from information in the Committed Model, with only the low level implementation-dependent detail coming from the Platform Model, specifically from the implementation of the firm objects.

8.1.1 The Two Stages of Transformation

It would be ideal if, through the application of Transformational Codesign, a single 'proven' model of a product could yield all the high level source code needed for its implementation. However, there are fundamental problems to be overcome by the transformation process that concern the communication between the soft and hard objects of the Executable Model and the ways in which hard objects are interconnected. These problems are the main reason that two models are needed.

We have seen that, in the Executable Model, implementation-specific features of objects are deliberately excluded and all communication between objects has a high-level language style. Regardless of whether objects are to become hard or soft, the lines connecting them represent events to be signalled or requests for data or actions. These effectively define the logical purpose of the objects and the connections between them. In an implemented system, the connections between soft and hard objects have to comply with physical reality and this means that they often have to follow a more tortuous path. For example, a hard object that is to service an interaction might be implemented with one or more externally accessible control registers, and these would appear in the address space of the software. In the physical hardware implementation, these registers will be interfaced on to the processor bus in a manner that takes account of processor and bus limitations, such as the operand size and other conventions of the address space. A typical interface protocol might require that the processor executes a sequence of instructions that manipulate these registers in a specified order to invoke the interaction and receive the result. Thus a place has to be found to document such instruction sequences and to associate them with the interaction they implement. It is not appropriate to embed this physical detail in the Committed Model, where it might escape some necessary attention when errors occur or changes are made. The form that the instruction sequence takes depends upon the detailed design of the hard objects, and so it is better to locate both in the same model.

If we look more closely into the implementation of the complete hardware of a computer system, we can see that, in addition to the hard objects from the Executable Model, a number of hardware components are needed, such as the processors and memories that provide the means to execute software, and the buses, arbiters and address decoders that need to be introduced to provide the necessary physical communication paths between the hardware objects. The hardware only comes together as a complete design when register interfaces of the hard objects have been designed and the physical connections between these objects and the processors, buses, memories, arbiters and decoders have been detailed. To continue the transformation of the Committed Model until it contains all the detail needed to synthesise a complete hardware implementation would not be compatible with the role and structure of a model based on logical behaviour.

In summary, the Transformational Codesign process produces two models and operates in two stages. In the first, transformations are applied to the Executable Model to produce a Committed Model, which shows by special markings the implementation decisions affecting the objects that contribute to logical behaviour. The second stage starts from an initial Platform Model, synthesised from the Committed Model containing all its hard objects, together with all the firm objects and including the interface objects that implement the software–hardware interface. Transformations to this Platform Model add the physical design detail for the hardware, the hardware-dependent implementations of the interface objects, and the firm objects.

The interface objects in the Platform Model service interactions that represent the high-level communication between the soft and hard objects. These are defined in terms of actions on the interface registers of the hard objects. At the same time, the hardware detail and physical interconnections are decided upon. Information flows between soft and hard objects are dealt with by the same mechanism since they can map directly on to interactions.

A similar treatment is required for event connections, although the role of the interface object in this case is rather different. For an event going from a hard object to a soft object, a typical implementation would be that the hard object causes an interrupt, as a result of which a function within the soft object is called to service the event. The transition from the hardware interrupt causing the software to be entered and a call to the appropriate event function involves the software in hardware-dependent detail, and issues of priority and pre-emption might arise. To localise the detail concerning the mapping of interrupts on to events, all interrupts for a processor are routed to a single interface object called the *Thread Manager* (see later), which calls the appropriate event functions.

The reader may have realised by now that connections between soft objects assigned to different processors will also pose problems. A solution to these problems is presented in Section 8.4, which is based on introducing hard objects to provide actual communication paths, and firm interface objects in each processor that exchange the inter-object traffic via a suitable protocol. The principle is similar to that used for making remote function calls on a computer network, but we must remember that we are dealing with objects rather than functions and, in embedded systems, we may not have access to the software packages that provide automatic support.

8.2 TRANSFORMATION OF THE EXECUTABLE MODEL

A variety of design decisions have to be taken during this first stage of Transformational Codesign, ranging from those that provide a suitable physical means for the external interface, to those that decide the assignment of soft objects to processors. Obviously it is also necessary to select appropriate hard or soft solutions for each component of the product, and these choices may consequently produce a need to make some adjustment to the style of the inter-object connections. As was noted earlier, this chapter concentrates on the kinds of decisions that have to be taken and the ways in which the Executable Model is transformed to incorporate the decisions. The criteria for making the decisions, which should be derived from Design Constraints, are not considered in detail.

The steps to be taken in transforming the Executable Model into a Committed Model can be summarised as:

- Implement external interface mechanisms.
- Commit objects to hard, soft or firm implementations.
- Decide on the number and type of processors and the objects associated with them.
- Analyse 'Threads of Execution' and resolve problems.

The principles involved in executing each of these steps are explained in the sections that follow. A complete example of their application can be seen in the Committed Model of the VCR in Appendix 3.

8.2.1 Implementation of the External Interface Mechanisms

The implementation of external interface mechanisms is the first step in Transformational Codesign since, in general, it requires the addition of new objects, which are best added to the model before other decisions are taken regarding the implementation techniques for specific objects. Thus each external connection to the model has to be examined to decide if its implementation requires the addition of objects that will provide the actual physical interface. Three kinds of external interface have to be considered separately. These are connections to external hardware, external computer systems, and human users.

External Hardware

Normally, connections to pre-defined external hardware will have already been designed to match the interface of a given hardware device, and these connections should not require further transformation at this time. The VCR model has several externals in this category; e.g., the Tape Drive, Aerial and Television shown on the External View of the VCR system. These can be seen in Fig. 5.2. In practice, interfaces of this kind correspond to wires that are to be connected to the finished product, and their behaviour will have been specified in terms of events and information flows. At a later stage of codesign, if these events and information flows are found to be connected to soft objects, then a hard object may have to be added to provide for software access to the connection. The same requirement applies to internal connections between soft and hard objects, and so transformations in this category are deferred until the commitment to hardware and software has been made.

External Computer Systems

The treatment of connections to external computer systems is an area in which it is difficult to generalise, and the actions to be taken depend on how the interface is modelled by the Executable Model. If the detail of an external connection is firmly specified at the start of a development, both logically and physically, the Behavioural Model should be constructed to conform to the specification of the hardware that implements the connection and hence the connection is in the same category as any other pre-defined hardware. However, if the design of a suitable interface with an external system is to be decided as part of the new development, the Executable Model should have an interface with the external system that conveys only the logical purpose of the connection. These are typically represented by interactions that establish a context in which the necessary detail can be evolved. During the first stage of codesign, an object is added to the model that represents the interface to be provided in the external computer system, and the external connections are routed through this object. Initially, the new object only needs to pass on the connection to retain the executable property of the model. In the second stage of codesign, i.e. when the Platform Model is developed, the detail for the connection will be developed by the same means that connections are developed for a multi-processor implementation of a stand-alone system. Since connection to an external computer system is not a feature of the VCR model, this issue will not be discussed further in this chapter.

Human–Computer Interface

Where a human user is to communicate directly with a computer system,

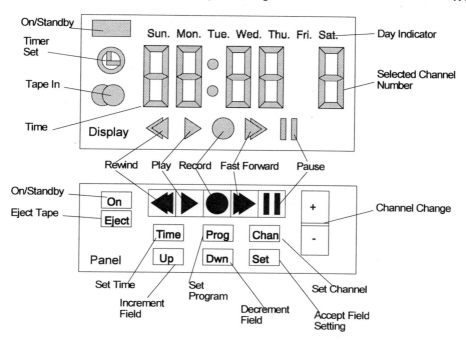

FIG. 8.1 The schematic of the VCR's user interface.

interface mechanisms are needed to allow the user to generate inputs and receive output. If the guidelines given in Chapter 5 have been followed, the Executable Model will be uncommitted with respect to the nature of these mechanisms, as it will show abstract logical connections with the user. For instance, the user of the VCR needs to provide the computer system controlling it with stimuli and information in the form of both event signals and a bundle of parameteric events (designated as **OTHER CONTROLS**).

During the first stage in codesign, the designer is expected to determine the physical aspects of how the system is to capture the stimuli and information, and should commit these decisions by transformations to the Executable Model. This commitment will usually involve the addition of new objects and associated changes to connections with other objects.

In general for computer system-based products, the options to be considered for input will include various types of specialised keypad, switches, key operated buttons, touch sensitive screens, pointing devices associated with displayed menus, and speech. Some products may be required to use more sophisticated devices, such as computer terminals, remote control devices and so on. For the VCR example, which is required to be simple and low cost, we choose to use a very simple panel of switches and buttons on the front of the VCR, as is shown schematically in Fig. 8.1. If the cost and simplicity constraints were not present, we would obviously consider remote control.

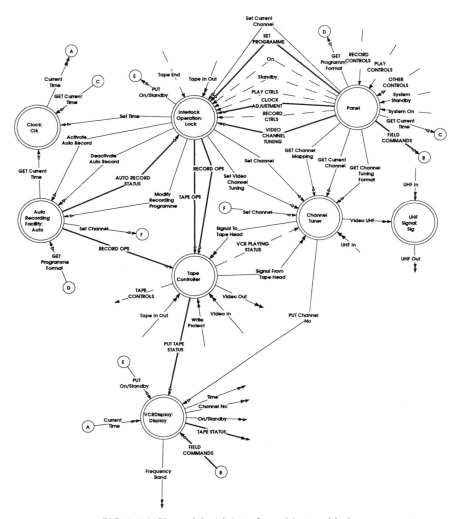

FIG. 8.2 VCR model with interface objects added.

There is also a need to select appropriate output devices and, in the case of the VCR example, the user is to be provided with the system status information. For example, there are several information flows directed to the user from objects such as **Clk** and **Channel Tuner**. Also, an output device is needed through which the system can prompt the user, to collect the parameters for the parameteric events concerned with setting the clock, setting up the channel tuning and specifying the automatic recording of selected television broadcasts. Due to the alphanumeric nature of both kinds of information, we shall assume that a single integrated gas plasma display is chosen to meet all user output requirements.

Thus the **Panel** and **Display** objects are added to the VCR model, as

Fig. 8.2 shows, and external connections to the user are re-routed through these new interface objects. The next step is to refine the parametric events by making explicit the design decisions on how each event is to be triggered and its parameters transferred. Since the overall policy is to minimise the spread of change, so that the changes may be properly regarded as transformations and the validated behaviour of the model is faithfully retained, the responsibility for obtaining the parameters is placed in the new interface objects. To achieve this placement of responsibility, some interactions have to be added between the **Panel** and **Display** objects, which can again be seen on Fig. 8.2. In practice, these new interface objects are not usually primitive, since they have both soft and hard parts that together will provide the required mechanisms and operating protocol. Thus new OIDs have to be added to specify the design detail, as can be seen in Appendix 3.

At this stage, as a result of re-routing the connections to the User, the **OTHER CONTROLS** bundle of parametric events will be connecting the **User** to the **Panel** and also the **Panel** to the **Lock,** as in Fig. 8.2. While the **User** connection to the **Panel** will remain unchanged, the one between **Panel** and **Lock** needs to be further transformed into an interaction having the same parameters. This is possible because interactions and parametric events are semantically equivalent except in one respect, namely that the interaction delivers all the parameters together with the stimulus to produce the action, whereas the parameters for a parametric event have to be collected after the stimulus arrives. A possible side effect of moving parametric events to different objects is that the original recipient might have been exercising its right to delay the response or even ignore the user who is triggering the parametric event, whereas interactions are expected to be serviced on demand.

In reviewing the impact of the changes made to parametric events on the VCR model, we observe that the **Lock** object has the responsibility of deciding whether a request – such as setting the clock – is to be allowed to start, depending upon the current operating state of the VCR. It also needs to know when the action is complete. However, the **Panel** object is to be made responsible for implementing the design policy with respect to obtaining the new value of the time.

Similar issues apply in the case of the other parametric events in **OTHER CONTROLS**, and the solution implemented in Fig. 8.2 is that each parametric event is replaced by a bundle of two interactions. For instance, the parametric event **Set Time** is transformed into the bundle **CLOCK ADJUSTMENT,** which is defined as the pair of interactions **Get Permission to Set Time** and **Set Time.** The intended design detail is that, if the **Lock** object authorises a time change, the implementation of the **Panel** object will display the current time and then take input from the 'Sel', '+' and '-' keys to obtain a new value of time, finally passing this new value to the **Lock** using **Set Time.** No further change is required and the roles thereby established for the objects have a stability that will allow revision of the time setting policy in a future product, at the cost of providing a new implementation of **Panel**.

8.2.2 Commitment of Objects to Hard, Soft or Firm Implementations

Although the criteria that govern the choice of a hard or soft implementation for objects are complex and varied, the feasible choices in many instances will be limited. For instance, we can see intuitively that, in the VCR system, the **Sig** object should be implemented in hardware as it processes television signals. In contrast, there is a temptation to conclude that the **Lock** object should be software, as it implements complex, context-dependent behaviour and does not appear to be under timing pressures. Thus a tentative assignment of objects to either hardware or software can be arrived at on an intuitive basis. The main issue then becomes more a matter of criteria to confirm choices, rather than rules for making them. We will need to evaluate what impact certain tentative choices have in terms of processor time, response time, silicon area, cost, memory requirements, etc.

It will normally be the case that, if the time to complete operations is tightly constrained, a hardware solution will be the preferred choice. Alternatively, if the availability of hardware that can be configured for special actions is limited, then a soft solution might be the best choice. However, the commitment of some objects may be more difficult, and careful analysis and experimentation may be required. It certainly cannot be concluded that anything that needs to be fast must be hard and anything that is very complex must be soft. Modern signal processors, coupled with the use of parallelism, can achieve very high performance rates with software algorithms and, similarly, hardware synthesised from high-level specifications can perform very complex functions.

Chapter 7 discussed techniques for predicting how constraints might be satisfied, and such techniques have an important role to play in codesign. However, the full consequences of commitment to hardware and software may only become fully apparent at the later stages of the Transformational Codesign process. Hence the commitment stage may have to be revisited and the commitments of some objects changed. The transformational nature of the approach and the provision of suitable tool support mean that work that has been carried out in the earlier stages may not have to be repeated for unchanged parts of the system, such as when others are subjected to changes of commitment.

We conclude this short discussion of how to decide between hardware and software solutions with the summary that techniques for prediction will assist the otherwise largely intuitive choice of hardware or software, and that choices that are very critical should be confirmed by simulation or analysis. The former is best for dynamic measures, such as perfor-

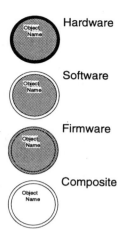

FIG. 8.3 Committed object markings.

mance, and the latter is best for the static measures of cost and size.

The commitment of objects to an implementation technology is shown by 'painting' the model so that objects are visually distinct, using the notation shown in Fig. 8.3. Its use can be seen in Fig. 8.4, which represents a fully 'painted' version of Fig. 8.2. A separate notation is not provided to distinguish between analogue and digital hardware. Consequently the interface is still abstract; for example, it is expressed in terms of information flows rather than physical wires, and the internal operation of objects is expressed in a procedural high-level language. Objects may be committed at any level of the hierarchy, the implication for non-primitive committed objects being that all their descendants are automatically committed in the same way. Objects with descendants that are committed to both implementation technologies retain the uncommitted representation; for example, the **Tape Controller** object in Fig. 8.4.

Figure 8.4 uses only two of the three committed object types. The third type, 'firm', is used to identify software objects whose implementations are expected to be hardware dependent (or possibly operating system dependent, as described in Chapter 9). No such example is found in the VCR model, but there is an example in the Mine Pump System in Appendix 3, whose Committed Model is developed as a Personal Computer (PC) with special hardware on to the PC's bus. In this system, there is a requirement to log the values of the gas concentrations, airflow, water level and pump status at periodic intervals. A printer was chosen as the logging device, so, when the External Interface mechanism was designed (by applying the advice of Section 8.2.1), an object **Printer** was added to the model. If the intention had been to implement the product as a stand-alone embedded system (with no operating system), this **Printer** object would have been marked as hardware. As it was assumed

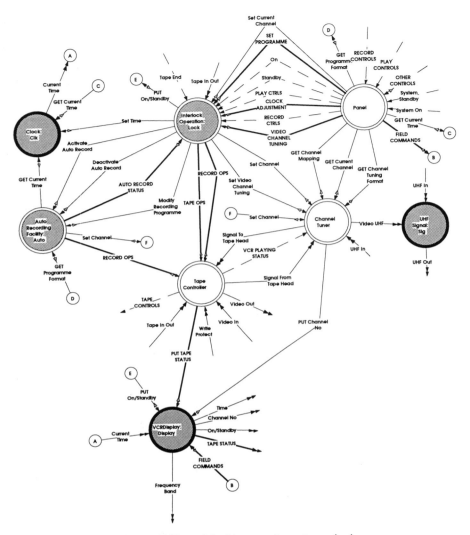

FIG. 8.4 VCR model with commitments marked.

that the PC's operating system would provide for printing, the **Printer** object was marked as firm, and so its implementation detail is not added until the Platform Model is developed.

The final stage of transformation associated with the commitment of objects to hardware, software and firmware is to transform the inter-object connections according to the rules set out in Table 8.1. This transformation is necessary because some of the inter-object connections of the model may be inappropriate to the implementation technology of the objects. For example, the most natural software-to-software communication is the interaction (corresponding to messages), and the use of infor-

TABLE 8.1 The transformation of inter-object connections after object commitment.

Connection From	Transformation of Interaction	Transformation of Event	Transformation of Information Flow
Software-to-Software	None	Interaction	Reverse Direction Make GET Interaction
Software-to-Hardware	None	Interaction	Make PUT Interaction
Hardware-to-Hardware	Information Flow(s) and Event	None	None
Hardware-to-Software	Reverse Direction and make GET Interaction AND Make PUT Interaction	None	Reverse Direction and Make GET Interaction

mation flows would probably be considered bad practice as it has the inference of shared data structures. Similarly, hardware-to-hardware connections are most natural in the form of events and information flows, and interactions are more difficult to implement due to the implications of synchronisation.

To avoid the problem that transformation of inter-object connections becomes a tedious and error-prone process, it needs to have tool support. These tools might offer suggestions for change, albeit in a stereotypical way, following defined naming conventions. Also, for some transformations – such as those for hardware-to-hardware connections – user input may be sought by prompting from the support tools.

8.2.3 Selection of Processors

At the point in Transformation Codesign that we have now reached, the model only contains the application-specific objects that are required to implement the system's functionality and its external interface. Although execution of this model can be simulated, as described in Chapter 6, the simulation depends upon objects being added automatically to the model to provide the execution environment for the application-specific objects. As the model is developed closer to a fully specified system, objects have to be added that will provide an independent execution environment. Principal among these are the processors that will *execute* the soft (and firm) objects, and will *drive* the hard objects that are to be connected to their buses. Thus decisions are needed at this stage of codesign regarding the number and type of processors that are to be used.

Like most of the other decisions made during the Transformational Codesign process, the selection of the type and number of processors will be strongly influenced by the product's Design Constraints. However, there may also be features inherent in an Executable Model that

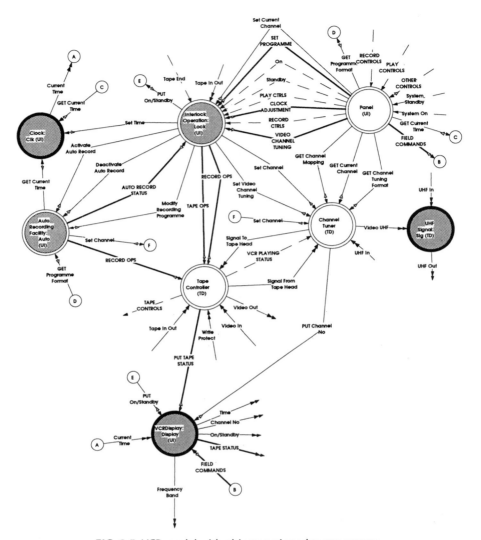

FIG. 8.5 VCR model with objects assigned to processors.

impose requirements for distribution or concurrency that are only achievable by using multiple processors. In the VCR example, we have chosen to have two processors, although this has been done to illustrate the treatment of multiple processor designs rather than because of actual needs. One (referred to as **UI**) is to handle the user interface and the other (**TD**) the video tape control.

Having selected the processors, they are not drawn as objects on the Committed Model. Instead, each primitive object in the model is associated with a processor by marking it with the name of that processor, as shown in Fig. 8.5. For software objects this marking defines that the operations of the object will be compiled to run on the named processor, and for hardware it implies that the object will interface either directly to the processor's bus or to other hardware connected to the bus. The assignment of objects to processors is again influenced by the product's Design Constraints, and careful analysis or modelling experiments may have to be undertaken before a satisfactory assignment is reached. In some circumstances, the decisions concerning processor allocation may need to be reviewed after an investigation of the overall system performance and behaviour has been carried out.

8.2.4 Analysis of Threads of Execution

Each processor provides only a single thread of execution, which is the execution mechanism for all the functions shown on the CIDs of the group of soft objects assigned to it. This is not ideal, since these functions may be concerned with servicing interactions or responding to events and, if they are not driven by these kinds of stimuli, they are assumed to be permanently active. In any object that has a mixture of function types, or even more than one event or active function, there is a concurrency implication. Thus some kind of quasi-concurrency has to be implemented for processors associated with such objects.

At this stage of codesign the precise nature of the quasi-concurrency provision may not be known. Even with an assumption that only bare processors are to be used, the policy of the Thread Manager object assigned to each processor may not yet be decided, particularly with respect to its use of interrupt mechanisms. If, later, the possibility of using a processor with an operating system kernel is to be considered, any provision it makes for multi-tasking will be relevant. Thus we conclude that it might not be possible to complete all the checks suggested below on the first pass, since some may need to be revisited as the Platform Model is refined. However, it is convenient to concentrate the discussion in one place.

Recalling the semantics of Executable Models from Chapter 6, the active functions are assumed to be executing continuously, and they are typically used to detect significant changes in information flows or internal state, or to complete tasks that need to run in parallel with other activities of the model due to the time they take to complete. Event functions are to be executed every time the associated events occur. The operation of all functions may produce activities in other objects by invoking their interactions. The synchronous nature of interactions implies that the execution thread can be passed from *calling* to *called* object, later to be returned with a result. In the light of this knowledge and in anticipation of the quasi-concurrent threads of execution that can pass through the model, an analysis should be carried out that investigates: the loading on each processor – for example, estimating the time taken to service events; reachability, such that the execution thread will reach each operation at the correct logical time; freedom from deadlock and livelock; and integrity – for example, that no two concurrent or quasi-concurrent threads can interfere with each other.

The quasi-concurrent behaviour of a Thread Manager may be based on different approaches in different applications. One typical design might use a processor's main thread of execution to run the active functions in a sequence that repeats indefinitely, rather than in parallel as the modelling notation implies. This tactic relies upon the assumption that, within each active function, there will be periods when there is no work to be done, in which case they would be 'busy waiting' for work if they had a dedicated processor. When they only have intermittent use of the processor, they must perform their checks, carry out any requested work and release the processor so that its thread of execution can be passed to the next function in the sequence. If a processor receives an interrupt that corresponds to an event, the main thread is suspended and a 'quasi-concurrent' thread of execution can be passed to the associated event function. Of course, event threads should not remain active for long, otherwise the time delays in servicing them and returning to the main thread may be problematic. Depending on the relative priorities of interrupts and the design of the Thread Manager, event threads may be capable of preempting each other, which adds to the complexity of the analysis.

Various transformations of the model may be needed if the analysis uncovers problems. It is not possible to generalise on these transformations, but examples include changing the processor assignments, shortening paths and introducing interlocking interactions. We will proceed on the assumption that no such problems occur in the VCR example, but integrity and interlocking are especially important issues and are discussed in general terms below.

Integrity and Interlocking of Critical Paths

The integrity issue is mainly one of checking that there are no concurrent paths through the model that can interfere with each other as a result of simultaneous access to shared data. Visual inspection of the CIDs in the context of which entries to an object might occur simultaneously is the basis of the analysis that is required. If the potential for interference is detected, some interlocking has to be provided. For example, the instruction sequences that have the potential to interfere with each other have to be regarded as critical paths that must be made mutually exclusive. In principle the solution requires that, on entry to a critical path, a check is made to determine whether or not any other object is currently executing a mutually exclusive path. This is easy to detect, provided that the check is applied at the start of all mutually exclusive paths and entry to a path is noted. However, the action required when a critical path is already active is not so straightforward.

Consider the case where the responsibility for the mutual exclusion of the paths is assigned to a new object, **Path Control**, that provides two interactions, say **Check** and **EndCheck**, to be used on entry to and exit from a critical path. By keeping appropriate state information, **Path Control** has the means for applying the check, but the action to be taken when the check fails is a problem. Logically the **Check** interaction needs to create the effect of 'busy waiting' until there is a call on **EndCheck** which frees the path. However, this would require that two interaction threads simultaneously exist inside **Path Control**, which violates a semantic rule of the modelling notation.

The solution adopted within an Executable Model is to provide a built-in class, **PathChecker**, whose implementation is integrated with the Thread Manager so that the context of a thread can be saved, other threads can be initiated and, when an **EndCheck** interaction occurs, the suspended thread can be continued.

When the Platform Model is synthesised, the objects in the **PathChecker** class will be carried forward as undefined firm objects, and suitable (implementation-dependent) implementations of the class will have to be produced. The definition of the class can take advantage of any assistance that might be given by an underlying operating system kernel and, failing that, it would make sense to provide implementations of these objects in the library for processors that are frequently used.

8.3 THE PLATFORM MODEL

As a result of the steps in the first phase of the Transformational Codesign process, a Committed Model is developed that contains the specification of all the functional objects, their implementation technologies and their association with processors. Further transformation is required before the model is ready to enter the final stages of the MOOSE paradigm, in which the implementation source code for the software and hardware is synthesised. However, the remaining transformations are confined to the completion of the hardware objects in the model, providing interfaces between software objects assigned to different processors, providing interfaces between the software and hardware objects, and providing the implementation of the 'kernel-like' firm objects. It is in order to focus attention on these objects that a Platform Model is constructed, containing all the objects that require further transformation. The MOOSE toolset automatically creates a Platform Model from the information contained in a Committed Model, which has previously been marked with processor allocations.

Initially the Platform Model contains a subset of objects present in the Committed Model, combined with the required synthesised interface

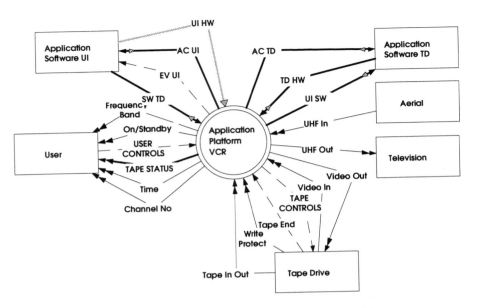

FIG. 8.6 The External View of the VCR platform model.

objects and some additional firm objects. It uses the OID notation and is structured in a similar hierarchical manner to the Committed Model. Thus an *External View* diagram provides the top-level view of the Platform Model. This shows the complete *Application Platform* as a single object. External objects are used on the top-level diagram as a means of defining and classifying the connections from the Committed Model that are to be supported by the Application Platform. First, all the externals from the External View of the Committed Model that have connections to hard objects are carried forward as externals of the Platform Model, together with the associated connections. In addition, for each processor referenced on the Committed Model, an external object is introduced on the Platform Model to represent all the soft objects assigned to the processor.

The connections shown between the externals and the Platform Model are bundled as indicated in the following list. Any bundles that would be empty are omitted. Since there may be several bundles of each kind associated with different processors, albeit with different components, they are distinguished by including the processor name in the bundle label, thus:

- Interactions from the soft objects to soft objects in another processor – **SW processor_name**
- Interactions to soft objects from soft objects in another processor – **processor_name SW**
- Interactions from the software of a processor to its hardware – **processor_name HW**

- Interactions from the software of a processor to its firmware – **processor_name FW**
- Event function calls into the software – **EV processor_name**
- Calls to active functions in the software – **AC processor_name**.

These conventions are demonstrated in Fig. 8.6, which is the External View of the Platform Model for the VCR.

8.3.1 The Structure of the Platform Model

The hierarchy of diagrams representing the Platform Model is designed to present views of the system that isolate different aspects of the transformational process concerned with completing the design. Thus the first level decomposition of the Application Platform provides the *System Communication View*. This shows a separate *Communication Object* for each processor that has soft objects that communicate with objects allocated to different processors. One responsibility of the Communication Objects is to translate outgoing (application-specific) interactions into *communication primitives* whose parameters convey both destination and parameter information in a general way. Objects are introduced later to route these primitives to corresponding Communication Objects in other processors. Therefore a second responsibility of the Communication Objects is to translate incoming communication primitives into the appropriate application-specific interactions directed to the soft objects allocated to their processor. They return responses through the reverse route. Since the Communication Objects have a stereotypical form their definitions can be synthesised.

The next level in the hierarchy, the *Software Interface View*, exposes the low-level software drivers required to interface to hardware objects, a Thread Manager for each processor, and the communication primitives. The interface objects receive the software-to-hardware interactions from the External View, and they have the responsibility for effecting these in terms of an appropriate set of transfers to and from registers in the hardware. Thread Managers are required to manage events emanating from hardware objects attached to their processor, and to schedule active functions.

Completing the hierarchy is a set of *Hardware Views*, one for each processor. Each of these contains an object representing the processor and all the hard objects from the Committed Model that are associated with it. At the time the Platform Model is synthesised, the interactions from the soft objects to hardware are routed through the interface objects to the appropriate hardware object. Subsequent transformation of the interface objects causes them to access the hardware through the processor bus, and the interactions in the hardware objects are transformed into information flows and events that produce the required effect. Thus the actual connections between software and hardware become very implementation-dependent, which is the reason this aspect of development is localised within the Platform Model.

8.4 TRANSFORMING THE PLATFORM MODEL

The ultimate goal of Transformational Codesign is to produce models from which implementation source code can be synthesised. When an Executable Model has been committed and its Platform Model has been synthesised, we are close to reaching the goal. The remaining steps refine the Platform Model, and they are concerned with the addition of hardware detail and the specification of the impact on low-level software. A summary of the required steps is:

- Design the register interface for the hard objects that interface to software.
- Develop the software interface objects.
- Implement inter-processor communication objects.
- Implement application-specific firmware objects.
- Develop the processor bus structures and add supporting objects.
- Implement the Thread Managers and review the integrity of the threads of execution.
- Provide hardware-level specification of the operation of the hard objects.

In more mechanistic terms, the steps listed involve adding new objects, predominantly but not exclusively of the hard variety, and transforming other objects so that they all come together as a physical system. For example, objects are provided to make the bus connections between processor pins and the functional hard objects. These in turn are provided with register interfaces, and software is written to manipulate the registers in a way that produces the required interactions.

Some interesting contrasts can be found between this phase of the MOOSE paradigm and the earlier phases in which the behavioural models are constructed. However, the theme of an integrated approach persists. In the early phases, the integration in the approach is realised by inventing collaborating objects purely on the basis of the need to cover behavioural responsibilities, and there is no distinction made between hardware and software. At the Platform Model stage, there is a very clear distinction between hardware and software, but the low-level hardware and software design still progresses in a very integrated way. This integration will be seen in the way that the steps listed above incrementally provide detail in the Platform Model, and dependencies are carried through between steps by the evolving model, even though some steps introduce hardware changes and others introduce software changes or possibly both.

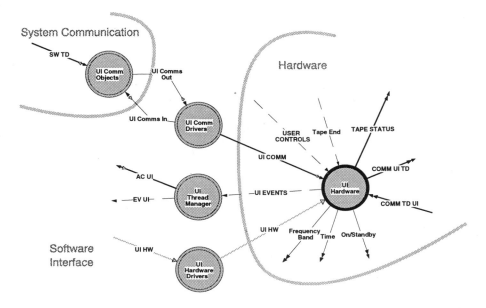

FIG. 8.7 A cross-section of the VCR platform model.

Figure 8.7 has been assembled from sections taken from the System Communication View and Software Interface View of the VCR Platform Model, to provide a background for the discussion of the transformations that are applied to a Platform Model.

8.4.1 Designing the Register Interfaces

The goal of this step is to design register interfaces for all hard objects that service interactions initiated by soft objects. The style of register interface provided by each object will depend upon the nature of the interactions that have to be serviced, and the physical characteristics of both the processor's data bus and theinstruction set.

Consider, for example, an object **Clk** contained within the **UI Hardware,** which has been assigned the responsibility of servicing the interaction **GET Time** in the bundle **UIHW.** This interaction is required to return the current day, hour and minute. Obviously, time could be encoded in 16 bits (3 for the day, 5 for the hour and 8 for the minute). The processor to which **Clk** is attached might, for other reasons, have a 32-bit data bus. Thus, if we were to use the full width of the data bus, it is more than adequate for mapping a single register containing the complete time value on to the address space. However, to illustrate a point, we might choose a bus width of only 8 bits. This then requires the **Clk** object to have at least two registers, but we might now go further and simplify the interface by having three registers, one for each of the time fields. These choices would result in an intermediate hardware view, in which the interactions entering the **Clk** object are replaced by *GET* interactions, each of which can be implemented by a single processor instruction. Assigning names that imply their implementation, the interactions might be named **GET8(day)**, **GET8(hour)** and **GET8(min)**. Obviously such interface transformations of the **Clk** object require corresponding changes to its class definition. However, this change can be achieved in a straightforward additive way by inheriting the required register interface from a new class definition. The entry for **GET Time** in the **UIHW** bundle would be replaced by the **GET8** interaction.

Both the user and receiver of the new interactions have to be further developed and, in the case of the receiving (hard) object, its register interface will be further developed when it is connected to the processor or system bus as described later .

8.4.2 Developing the Software Interface Objects

After the decisions have been taken concerning the form of the register interfaces of hard objects, the functions of the interface objects in the Platform Model can now be coded. The operations provided in the interfaces of these objects are what the higher-level software expects, and implementations have to be provided that map each interaction on to the appropriate set of reads (GETs) and writes (PUTs) to the relevant registers. This process is very straightforward, as can be seen in the example below which demonstrates mapping the **GET time** interaction on to the **Clk** object's register interface.

```
TimeType Get_time (void){
   TimeType temp;
   int day, hour, minute;

   temp.day = (int) CALL (GET8 (day));
   temp.hour = (int) CALL (GET8 (hour));
   temp.minute = (int) CALL (GET8 (min));
   return (temp);
}
```

In the jargon of C++, functions such as the one shown above may be defined as in-line functions, in which case the given code sequence will be substituted for every call of **Get_Time,** no function calling overhead need be incurred, and there will be no instantiation of the interface object. However, its presence on the Platform Model is still justified because it places the responsibility for defining the detail for accessing a hard object with those who design the hard object's interface.

8.4.3 Implementing Inter-Processor Communications

The problem here is to implement the communication primitives that are serviced by the **UI Comm Drivers** objects. These objects will divide into a firm object to implement the primitives and hard objects to effect the physical connection. The latter should be added to the Hardware Views of each processor, and the **UI Comm Drivers** can be developed into an implementation of the software interface using the registers of the hardware as described earlier. It might normally be expected that library objects are provided which use standard communication mechanisms or, possibly, shared memory. If the development of new components is required, the hardware interface and the soft interface objects are developed in the same way as the connections between the soft and hard objects described above.

8.4.4 Implementing the Application-Specific Firmware Objects

In general, firm objects are those that have hardware dependent implementations. We have already discussed several kinds of firm object, of which the simplest are those providing an interface between hardware and software. The most complex is the Thread Manager, that handles interrupts and processor scheduling in a manner that we have not yet described. All of these firm objects are automatically placed on the Platform Model by the synthesis process.

As noted earlier, firm objects may appear in the Committed Model, the implementation of which depend upon detail to be decided in the Platform Model. There are no examples of this in the case of the VCR model but, if there were, an additional object would appear in the Software Interface View, which would contain all the objects marked as being firm. They are implemented like other soft objects, although the implementations make use of physical detail available only to the designer developing the Platform Model.

8.4.5 Developing Bus Structure and Adding Supporting Hard Objects

Transformations at the previous stage have resulted in hardware objects with register interfaces that can be mapped directly on to the processor's bus. However, a set of supporting hard objects may have to be added to the model to complete the bus connections between the processors and the register interfaces of the hard objects. This step is where the memory space is allocated and the memories associated with processors are chosen, so that objects are added to the model to represent the memories and provide address decoding logic. Since the bus is also the vehicle that will convey interrupt requests to the processor, objects are introduced at this stage to map the hardware-generated events on to the interrupt structure of the processor. The addition of these objects requires transformations to the Hardware View that replace interactions by data buses and signals, which make use of both events and time-continuous information flows. Thus the **GET8** interactions into the **Clk** object might be transformed into a connection to the processor's data bus and a set of events that emanate from the address decoder. From this transformation, an address map is constructed that is needed by the firm interface objects that access registers.

8.4.6 Implementation of the Thread Manager

A Thread Manager is required for each processor. Its specification is largely derived from the analysis of the thread of control conducted at an earlier stage. At this point, a class definition has to be produced for the Thread Manager of each processor, the detail of which is dependent on the requirement imposed by the application-specific soft objects, the processor and other hardware detail. Consequently its implementation is difficult to describe, and only a summary of its features will be given. There is further discussion of this topic in Chapter 9.

Thread Managers receive all interrupts from hardware objects and map them on to calls to the event functions of the objects. This responsibility is evident in the connections entering and leaving a Thread Manager. They are also responsible for calling active functions. Thus the Thread Manager might be regarded as the scheduler of processor time. When an application is started, the processor's thread of execution is passed to the Thread Manager. It is also required to have a constructor function, to make appropriate entries in the interrupt vector table of the selected processor and subsequently to provide interrupt-driven code that generates the outgoing event calls, according to the decisions regarding the mapping of events on to interrupts. The Thread Manager may also be required to support the **PathChecker** function described in Section 8.2.4.

8.4.7 The Provision of Hardware Level Specifications for the Hard Objects

The behaviour of all objects in the Platform Model is specified initially in the same way as that in the Committed Model, which, in turn, is derived almost directly from the Executable Model. Thus hard objects that come through this route have their behaviour specified as a graphical CID with FSPECs written in C++. Furthermore, this C++ code is written before the objects are committed to hard or soft implementations. Although, in the case of soft objects, the C++ code can satisfactorily be re-used as the implementation, in the case of hard objects it is best to regard it as a soft emulation of the object. To argue that, with current techniques, an acceptable implementation could be synthesised from the C++ emulation would be rather more ambitious than prudent.

For the hard objects whose implementations are to be bought in, either in the form of 'off-the-shelf' ICs or library cells available in a particular foundry process, synthesis of an implementation is not an issue and the emulation in C++ may be the only detailed specification that is required, provided that the source code of the off-the-shelf implementation is noted somewhere in the model. For those objects for which a custom solution is required, a new class definition has to be produced in a language from which hardware synthesis is feasible. MOOSE does not firmly specify what the synthesisable language should be, and the policy is to go with whatever the industrial strength hardware tools provide. At the time of writing, this is VHDL, since tools are available to carry out electrical level simulation and a silicon compilation for a system specified in VHDL.

The simulation facility in a VHDL tool set may make it worthwhile to provide a VHDL specification, even for the bought-in components of a system. In fact, for library objects that correspond to bought-in technology, both C++ (soft) emulation and VHDL (hard) emulation would be appropriate, although for commercial reasons it is more than likely that the VHDL model will be written in a high-level behavioural VHDL rather than a lower-level VHDL from which the implementation could be synthesised.

In all cases the producer of the VHDL definition of an object should treat its soft emulator as a firm (and validated) behavioural specification, and should maintain a compatible interface specification. In fact the same CIS that was used for the soft emulator can be used, provided that alternative type definitions are added that give VHDL type information. For example, an analogue-to-digital converter, which has information flows to represent an analogue input X, a digital output Y and an event (strobe) E, would have the following CIS:

DATA: X T1, Y T2;
EVENT: E;
CTYPE: T1 {float}, T2 {int};
VTYPE: T1 {real}, T2 {sig [16]};

The other two components of the class definition, the CID and the function specifications, should be written so as to define the object as a VHDL entity, with functions that correspond to VHDL processes that service the interface.

8.5 SYNTHESISING AN IMPLEMENTATION

On completion of the Transformational Codesign, outputs can be synthesised from the Committed and Platform Models that specify in VHDL the

ASICs required to build the product, the net-list describing how the ICs are to be connected, and a stand-alone program for each processor in C++.

It is the **UI Hardware** object shown in Fig. 8.7 that is used to synthesise the VHDL and net-list, which define the hardware of the system. If it is required, a single VHDL model can be produced for the total hardware system and, in principle, a single chip solution could be manufactured from it. Of course, the VHDL is further processed by standard tools used in the manufacture of ASICs. In many cases, some objects, such as the processor, memory and interrupt handlers, will be taken from libraries, and these may not be described fully in VHDL. These objects will not be synthesised, but it is assumed that appropriate layouts exist on to which the other objects can connect. Pragmatic issues concerned with using VHDL and net-lists are discussed in Chapter 9.

For analogue hardware, for which the synthesis of implementations from design notations is not well established, the Committed Model provides a detailed behavioural specification from which the analogue parts will be engineered in a conventional manner. Conceptually, these objects contain the analogue hardware and provide it with a digital interface.

The information needed for automatic generation of the C++ programs is divided between the Committed Model and the Platform Model. Both models are constructed as hierarchies of OIDs and, in both cases, class definitions should exist for every primitive object in the model. Further, in each model the software objects are marked with the name of the processor on which they are to execute, which is equivalent to having a separate hierarchy for the objects in each processor.

The class definitions of the primitive objects in MOOSE are easily translated into C++ classes and, although class definitions for the non-primitive objects in the model are not strictly necessary, there is enough information in the diagram hierarchy of the Committed Model also to produce them automatically. Hence C++ class definitions are automatically produced for the objects that are to execute in each processor which reproduce the object hierarchy specified by the Committed Model. Therefore all the application-specific objects to be executed by a particular processor can be placed in its 'program' by declaring an instance of the object at the top of its hierarchy. Objects that depend upon the hardware detail have classes defined in the Platform Model. Some of these may be provided as in-line functions, hence the code is used but objects will not be instantiated. Others, such as the Thread Manager, will be instantiated in the Synthesised Program.

9 Pragmatics of Using MOOSE

In presenting the MOOSE paradigm for computer system development, we have dealt with a notation and method for producing an architecture for a computer system product, and the commitment of this architecture to a specific implementation through a process termed Transformational Codesign. Obviously the application of the paradigm through to the stage where a tested product exists requires an intimate knowledge of the implementation technologies. Coverage of this is beyond the scope of this book. In the interest of clarifying the principles of the MOOSE approach, earlier chapters have simplified or omitted a number of pragmatic aspects of producing an implementation. This chapter reviews the main areas in which some important pragmatic considerations were passed over, and it discusses the implications for the application of MOOSE to industrial strength product developments. The issues to be discussed fall into four disparate categories. These are:

- The use of standard system software to support the application-specific software.
- The physical construction and packaging of hardware.
- The implementation of hardware.
- Aspects of simulation that relate to the evaluation of the performance-critical decisions that are taken during the codesign phase.

They all affect activities late in a development when the specialised skills dominate, and they may not all relate to the area of specialisation of some readers. However, all are of importance to the successful creation of industrial strength products, and the skills of computer systems engineers should cross such boundaries, hence they have a place in this chapter. The separation of the topics into distinct sections, Sections 9.1 to 9.4, will allow the reader to be selective, and each section has been given an introduction intended for those without specialist knowledge in that area.

9.1 THE USE OF STANDARD SYSTEM SOFTWARE

The principal and most basic component of system software is the operating system. Among its many roles, an operating system insulates an application program from hardware detail and generally makes a computer easier to use. However, the VCR example given in Chapter 8 has demonstrated that, in some applications, it is practicable to implement the application-specific software for the bare processor, even though we must concede that this is not normal practice, except for applications in which the processors are too primitive to support a useful operating system or where special features of the application make the use of a standard operating system inappropriate.

We now need to consider carefully the case for incorporating an operating system into a system developed with MOOSE, and to discuss how this impacts upon the process of transforming a model into an implementation. Since for many applications the MOOSE approach makes it feasible to use bare processors, it would be inappropriate to accept without question the popular view that it is becoming too difficult to implement application software for a bare processor. We will set the context for the discussion by summarising the features and identifying the role and functions of operating systems, before addressing specific issues pertaining to the case for choosing to use an operating system. Finally, the pragmatics of integrating the application software with an operating system are described.

9.1.1 The Role of an Operating System

A general purpose operating system provides all the facilities needed to develop and execute an arbitrary collection of application programs. Typically it will have specialised facilities to aid program development, and a broad range of built-in functions (often called supervisor calls) to meet the run-time needs of these programs, such as allocation of memory and management of input and output. Depending on its level of sophistication, an operating system may also supply the application software with an idealised view of the computer; for example, by providing a virtual memory of almost unlimited size, and by the *multiprogramming* of several programs to create the impression that they are running concurrently, while protecting these programs from accidentally interfering with each other. Other major facilities, such as a file store, database system or an interactive (perhaps window-based) interface with the user may also be available through supervisor calls.

The multiprogramming feature is only one example of how an operating system can provide an illusion of concurrency. Even more relevant to the implementation of MOOSE models is a *multi-tasking* facility in which programs can specify tasks within themselves that are to appear to run concurrently. In practice, they will not be concurrent in the way that tasks running in separate processors would be, and so a distinction is made by using the term 'quasi-concurrent'. Also, tasks are not usually independent in the way that programs are, and a multi-tasking operating system would provide supervisor calls to facilitate inter-task communication and synchronisation. For instance, tasks which access shared data will require facilities to request exclusive access to the data if a sequence of interrelated changes need to be completed before another task is granted access.

Computer systems will not normally require the rich range of facilities found in a general purpose operating system. They will be more concerned that the performance of the application-specific software is not impaired by system overheads and that response times to external events are predictable. Thus *real-time* operating systems are provided, which have reduced facilities but better multi-tasking performance and predictable response times. Since application-specific computer systems only need to run a fixed set of software, facilities for program development can be omitted and a minimal set of supervisor calls, including those concerned with multi-tasking and the interfaces to external devices, may suffice.

The above description has used the term 'supervisor call' loosely, and some clarification is needed. Normally the term is used for the functions built in to the central part of an operating system, the *kernel*, and they are entered by a special mechanism that crosses a protection boundary. For example, a supervisor call might be made by an instruction that causes an interrupt (or 'exception'), which causes the processor status to change in a way that allows the kernel to access parts of the memory protected from the application-level software. Normally only the basic essential functionality, which needs to be implemented by functions executing in the privileged ('supervisor mode') of the processor, is provided by actual supervisor calls. A higher-level set of functions, which themselves make use of the 'kernel functions', can provide a more sophisticated interface for the application programs. However, for many computer system applications, the kernel may be all that is required or indeed can be afforded. Thus we shall use the term kernel in what follows to mean a reduced-facility operating system, without meaning that its scope is limited by physical protection issues. The functions it provides will be referred to as supervisor calls, without distinguishing between privileged and non-privileged functions.

If a multi-tasking kernel is used to implement a MOOSE model, it is likely that the system developer will need to be aware of the ways in which the quasi-concurrency it provides differs from the real concurrency provided by separate processors. Clearly, when there is only one processor, only one instruction can be obeyed at a time. This may belong to one of the tasks of the application software, to a supervisor call function in the kernel, or to an interrupt servicing function of the kernel. Except when the processor is executing instructions at the highest priority level, there is the possibility that, on completion of each instruction, the processor might be pre-empted and allocated to a different activity. When this occurs, a considerable time could elapse before the next instruction in sequence is obeyed. It is hardware interrupts that pre-empt the processor in this way and it is the kernel that decides what the processor should do next.

In general, at the end of each excursion of the processor's thread of execution through the kernel, whether this is the result of a supervisor call or an interrupt, the kernel again has the opportunity to decide on the basis of its scheduling policy whether it should return the thread to the current activity or temporarily suspend this activity and effect a task change. It is these *context changes* between different activities that produce the quasi-concurrency, and this only meets real concurrency needs if the time elapsing before suspended activities are resumed has no noticeable effect on overall behaviour.

To conclude, we summarise the features required in a kernel (or operating system) that is to support computer system applications. These applications must be able to respond very quickly to events, often signalled by interrupts, and the responses need to come from the application-dependent software. Events must not fail to be serviced, even when the system is under the pressure of a peak workload, and the responses may have to meet strict time deadlines. Thus the speed of interrupt handling and context switching are critical to achieving correct operation. Also, mechanisms have to be provided for communication and synchronisation between different tasks within the application, since these will be working together towards common goals. The requirement for other kinds of support of the kind provided by general purpose operating systems is naturally variable and, in each particular case, is limited. Therefore a modular structure that allows only the essential features to be selected is attractive.

Of course, general purpose operating systems have many features in common with real-time kernels, but they are typically much larger, are not subjected to the same tight performance constraints, and often have recovery or avoidance paths not open to the real-time system. For example, if a keystroke from a terminal is occasionally missed, the user will instinctively strike the key again. Also, sophisticated input/output buffering techniques can be used so that the immediate actions needed when interrupts occur need not involve application-specific code. Responses in general can therefore be slower, since they often only need to match human expectations and not the demands of a potentially dangerous process that the application software is controlling. Another difference that eases the pressure on a general purpose operating system is that the programs it runs are usually independent of each other, hence it can concentrate on the protection of one activity from another rather than the coordination of their interactions.

9.1.2 The Case for Choosing an Operating System

The user of a general purpose operating system would normally form a view of the system from the functionality it offers to application programs, the way in which it manages the user interface, and the general purpose applications that can be pre-loaded. The case for choosing a particular system would be based on this view, and it is wholly unrealistic for such a user to contemplate using the computer without an operating system. In contrast, the computer system developer has to be more concerned with the cost and overheads of the functionality offered, particularly if some of it is not strictly necessary, and with the flexibility of its input and output (or 'device driving') features. It is unlikely that a full general purpose operating system would be considered unless this was part of the specification; for example, in the processing nodes on a very large distributed system.

More likely, the realistic choices available to the computer system developer will be either a kernel offering limited capability, a modular kernel that can be customised, or a fully customised solution implemented for the bare processors of the Platform Model, as described in Chapter 8. In general there may be several processors in the system, and hence in the Platform Model, and different choices may be made for each of them. Once the choices have been made, the kernels should be regarded as integral parts of the processors on which they operate. Obviously the supervisor calls they provide can be used in the code that has to be written for the functions of the soft objects of the Platform Model.

In the MOOSE approach, the existence of a kernel affects only a limited and well defined part of the total software of a system, and this is confined to the Platform Model. For example, the software in the Committed Model will be kernel-independent, although it might require kernel type facilities to be provided in the Platform Model (mostly in terms of the objects that are committed as *firm* – see Chapter 8). Regardless of the choice of kernel, the external specifications of the firm objects will have to be met by the implementations developed for them in the Platform Model, even if suitable supervisor calls are not available. Thus the benefits of a kernel can be assessed largely in terms of the impact that its supervisor calls make on the problem of implementing a relatively small number of well specified functions.

There are two additional benefits in using a kernel, beyond the functionality provided by its supervisor calls. First, the facilities for interrupt handling and multi-tasking might be useful in the implementation of the concurrency-related aspects of the 'Thread Manager' objects in the Platform Model. Associated with this are the benefits of having the kernel deal with the synchronisation that might be needed by any functions that share data. Second, interprocessor communication might make use of the kernel facilities together with off the shelf hardware. This is of significant value in a system that has to be physically distributed and where connections can be made through standard mechanisms such as Ethernet.

It could be argued that another benefit arises from the 'virtual processor' provided by a kernel making the application software processor-independent and portable. These are obvious attractions and, in the MOOSE approach, they have been a fundamental objective behind the entire paradigm. Thus processor-dependent features are localised and well encapsulated inside a Platform Model, but the abstraction in the models that exists from the architecture level downwards is even more important. This applies also to the libraries where objects with clean functional interfaces can have multiple implementations targeting a variety of technologies. Therefore the authors do not regard portability or 'openness' to be a substantive case for using a kernel although it can make a contribution.

In summarising this discussion, we conclude that the case for using a kernel will hinge on the kernel providing the best means for implementing the functionality required in the firm objects, which includes the Thread Manager and Communication Objects. However, the following questions have also to be addressed in order to decide what is 'best': 'Does the use of a kernel introduce new problems for the developer?'; and, 'Are there features of the Platform Model that cannot be implemented without kernel assistance?'

Does the Use of an Operating System or Kernel Cause Problems?

Although the use of an operating system or kernel can simplify the implementation of a computer system, there are a number of problems which might arise from their use. The first is simply an issue of size; operating systems are not small. For example, a basic DOS system on a PC will typically require about 15 Kbytes of RAM and approximately 60 Kbytes of disc space. These requirements are substantially increased when utilities are added to the basic system. For example, operation through a Microsoft Windows™ interface requires a minimum of 1 Megabyte of RAM and 20 Megabytes of disc space. A more sophisticated operating system, such as UNIX, will raise the requirements even higher. However, as indicated above, computer systems would not use these general purpose operating systems unless the requirements specified the need, for example, to provide processing nodes in the form of industry standard workstations. It would be more likely that a real-time operating system or a micro kernel such as Chorus (Rozier *et al.* 1988) would be under consideration. Even so, in the case of low cost and physically small products, the overheads may be significant. For example, a small kernel might require 5 Kbytes; a more typical size would be 30–50 Kbytes.

Although we often hear that memory costs are low and constantly decreasing, these claims need to be put into perspective. Whatever the cost, the impact may be more or less severe according to the purpose of a product development. For instance, many 'mass market' computer system-based products fall into the $100 to $1,000 price range. These products involve relatively high packaging and marketing costs, hence realistic manufacturing costs for the computer system component may need to be limited to the $10 to $100 cost band. Clearly such stringent cost constraints dictate that system resources are kept to the absolute minimum necessary to deliver the required functionality, and nothing more than a minimal kernel could be considered. Even when price constraints are less severe, the cost associated with using a kernel needs to be evaluated against the tangible benefits that accrue, and clearly these will depend upon the extent to which the facilities of the chosen system are needed. In the MOOSE approach, the use of kernel facilities is confined to the firm objects in the Platform Model, which facilitates the assessment of their benefits.

Another aspect of the context in which computer system developers work is that the systems are subject to real-time constraints; for example, throughput requirements in the case of input and output, and completion deadlines in the case of the functions that provide responses to external stimuli. The kernel represents an execution time overhead that must be taken into account in any assessment of a system's ability to meet its time deadlines. Sources of overhead include the execution time for kernel code, interrupt latency due to servicing higher priority activities, the time for which tasks may block each other's execution through critical path synchronisation, and the time taken to effect context switches. Although most developers of real-time kernels have expended considerable effort in minimising these overheads, they still remain, and they may be difficult to quantify unless the kernel source code is available.

If a multi-tasking kernel is to be used, the issue of scheduling has to be addressed. That is, given an estimate of the task execution times, the kernel overheads and the kernel's scheduling policy, is it possible to schedule the tasks for execution in such a manner as to guarantee that the system's time deadlines will be met? Such an analysis is usually difficult, and it becomes even more complex if the tasks can experience blocking during communication or synchronisation (Burns 1991). Recent progress in dealing with these issues is discussed by Audsley *et al.* (1993) and Burns *et al.* (1995).

The use of standard, commercially available real-time kernels may lead to other difficulties. For example, many kernels in this category use fixed priority pre-emptive scheduling, and this can lead to the phenomenon of priority inversion, where high priority tasks are effectively descheduled to make way for low priority tasks. A number of priority inheritance protocols have been proposed in the literature (Sha, Rajkumar and Lehoczky 1990) which can avoid this problem, and some also guarantee the maximum number of times per activation that a high priority task sharing a resource with lower priority tasks may be blocked waiting for the resource. However, in using a real-time kernel, the computer system developer must be aware of the potential for inversion and take steps to use an appropriate kernel if priority inversion is deemed likely to compromise the system's ability to meet its time deadlines.

The degree of redundancy in a kernel has also to be assessed. Although modest redundancy can be accepted almost without question in general purpose computing, it is to be avoided wherever possible in computer systems. Exceptions would be if the redundancy has a planned purpose, such as improving reliability, or its presence is a consequence of reducing development risks, such as those concerned with time-to-market goals. In fact, in the computer systems that are to be implemented entirely on silicon (the so-called deeply embedded systems), the tolerance level for any form of redundancy is very low since silicon area is at a premium and manufacturing and testing problems are adversely affected.

In such an extremely cost-conscious context, a modular operating system kernel is a possible solution – assuming that an optimum selection of modules can be made – which eliminates all redundancy. However, it is not intuitively clear that modular kernels can achieve the zero redundancy goal. Provided that extensive operating system features, such as file systems and networking provisions, are not required, the MOOSE approach provides a possible 'do-it-yourself' alternative by making it much more feasible to use a bare machine, as was demonstrated in Chapter 8. The exploitation of re-use in implementing a Platform Model, which can be achieved through library objects, further reduces risk and de-skills the work.

Yet another issue to be addressed is the extent to which the application requires completely deterministic behaviour. Currently, the use of concurrency or quasi-concurrency is prohibited in a number of safety-critical (and other high integrity) systems, such as certain avionics applications. For systems of this kind, cyclic scheduling is usually used to provide the desired degree of predictability, although Burns et al. (1994) argue that scheduling theory for real-time systems is now sufficiently well developed to allow restricted forms of concurrency or quasi-concurrency to be used in safety-critical systems when supported by a safety kernel. Such kernels are still the topic of research. They provide mechanisms to ensure that tasks execute free from unplanned interference and ensure that, at all times, the system is maintained in a safe state.

The discussion in this section has cited examples of issues that need to be considered in deciding if an operating system or kernel is an appropriate choice for a particular computer system product. It has not provided a complete checklist, because each product development will provide its own interrelated constraints, such as limited choice due to lack of skills or experience. Others issues will depend upon the details of an implementation. Consider, for example, the case where the 'best' implementation of a system, for performance reasons, is one that has some objects implemented in hardware and others in software. A kernel, indeed even a modular kernel, may impose unacceptable constraints on the way in which these hardware and software objects can communicate, whereas the throughput and response requirements for the system may require that the communication is fully customised without the inclusion of any operating system enforced protocol.

Does the Absence of Kernel Facilities Pose Problems?

The answer to the above question is a qualified 'yes'. For example, if the behaviour to be provided by a computer system requires it to have a conventional file system, then it is unlikely that any solution that avoids using the file system of an existing operating system would be successful. This is not to say that, if the requirement is simply to store information on disc, a custom solution is infeasible. Similarly, if an application demands that the hardware base for its implementation takes the form of a distributed network of user stations, the best choice of implementation platform will probably be industry standard workstations, together with their standard operating systems, connected by standard network technology such as Ethernet. In contrast, if the requirement is for limited communication between two processors with access to a common bus, an operating system would be 'overkill', and a solution could be engineered in a customised way by using shared memory on the bus. Clearly, if an approach such as MOOSE is to be acceptable across a wide range of systems, it must provide the implementor with choices as wide ranging as those mentioned, and assist in their evaluation, to achieve the best match with requirements.

Multi-tasking is another commonly required kernel facility. The 'bare machine' treatment, presented in Chapter 8, of a computer system involving concurrent objects, required the implementor to determine and implement the policy for sharing a processor among the active functions and event functions of its objects. When simple scheduling rules suffice, implementing them in a Thread Manager is not a serious problem, but the level of difficulty increases with the complexity of the scheduling rules. For instance, if pre-emption of the main thread of execution is required while it is executing in a function other than the Thread Manager, the full context changing mechanism of a multi-tasking kernel would need to be reproduced, and the complexity that this entails may make any other costs associated with accepting an operating system kernel solution quite tolerable.

We may therefore conclude that, to provide a computer system with the functional capability available in operating systems, such as file and communication facilities, and to treat concurrently executing objects as concurrent tasks, serious consideration should be given to implementing a MOOSE model by using a commercially available operating system, a kernel or a modular kernel that can be configured to suit the application. In the next section we examine how the codesign process described in Chapter 8 needs to be modified to integrate the application-specific software with such systems. Another conclusion is that, when the needs of the application are more basic and the appropriate skills and experience are available, it may be better to take on the challenge of making the application run on the bare processors. If this is done in the recommended style, over a succession of projects, a flexible and modular operating system kernel will evolve in the form of re-useable objects. The final conclusion is that a development has to be completed on time and with the skills available to meet time-to-market constraints. It may not be possible to deliver on time without importing some form of operating system and possibly other utilities, in which case there is definitely no alternative, whatever the cost.

9.1.3 Integrating a MOOSE Model with an Operating System Kernel

We have seen that the decision to use a kernel in any (or all) processor(s) of the implementation affects only the implementations of certain objects of the Platform Model, such as the Thread Manager and other firm objects that provide operations similar to the supervisor calls found in a kernel. Due to this localisation of the interface between an application and a kernel, it is a relatively straightforward task in MOOSE to integrate the application-specific software with a kernel. However, having done so, it may be necessary in the case of some systems to re-assess the overheads.

A simple Thread Manager for a bare processor was discussed in Chapter 8, where it was suggested that the processor's main thread of execution should be used to execute the active functions in an indefinitely repeating sequence, and that this thread should be interruptible. Interrupts can be exploited by the Thread Manager as the means for breaking into the continuous execution of active functions, so that events can be detected and the associated event functions called. The use of interrupts to trigger events does not demand a direct one-to-one mapping from interrupts to events, and in the extreme case, one periodic interrupt into the Thread Manager would suffice, with an associated interrupt action that polls registers in the hardware to discover the occurrence of events.

Naturally this use of interrupts requires the Thread Manager to make appropriate settings in the processor's interrupt vectors. Its detailed design must also determine the interrupt priorities and whether or not event functions are allowed to pre-empt the processor when it is executing other event functions. Those familiar with the interrupt entry mechanisms of typical processors will realise that the Thread Manager might also need to manipulate hardware registers at the point of interrupt entry to secure the register values of the interrupted function and make it possible for new function calls to be made using the normal stack and return mechanisms.

A Thread Manager based on a kernel could operate in much the same manner, but it would not need to be concerned with so much processor-level detail, such as managing register dumps. Some kernels might allow the users, which in the MOOSE context means the developer of a Thread Manager, to compile their own device drivers into the kernel. Normally in MOOSE, the 'device driving' for the application-specific hard objects remains the responsibility of the interface objects in the Platform Model, and it is not transferred to either the interrupt entry sequences in the Thread Manager or the device drivers added to the kernel.

The use of a multi-tasking kernel can have a more significant effect on the Thread Manager since, in this case, active functions and events can run as separate quasi-concurrent tasks. As indicated earlier, the tactics adopted by this kind of Thread Manager would have to be known when the final analysis of threads of execution is carried out, and additional firm objects might have to be introduced to implement interlocking mechanisms for critical paths.

In spite of a superficial attraction in offloading the responsibility for managing the concurrency on to a kernel, we do see this as its main attraction. The greatest benefit of using a kernel lies in the supervisor calls that are used in the implementation of those firm objects that are providing operating system type functionality, such as specialised interlocking functions (for example, semaphores), a file system, a standard networking provision and input/output management. As we have already noted, if facilities of this kind are needed and they are not provided by the kernel, then they will have to be provided the hard way.

9.2 THE PHYSICAL CONSTRUCTION AND PACKAGING OF HARDWARE

Our second major area of discussion concerns the production of the hardware platform on which the software of a computer system is to run. A basic assumption with respect to the physical construction of this hardware is that it will be packaged in the form of Integrated Circuits (ICs) mounted on Printed Circuit Boards (PCBs). A large system will require several PCBs, which might be mounted in close proximity in a card cage, and the cages might be physically distributed and connected by means of wires or optical fibres. Even more widely distributed systems will use 'wireless' communications in a variety of frequency bands. The cages in which the boards are mounted will usually have standard backplanes, which provide direct connections between the boards. These cages might be stand-alone or an integral part of a computer, such as an industry standard workstation whose processor is able to access the boards via the backplane, hence it may act as one of the processors in the system.

A single MOOSE model is sufficient, and indeed desirable, as a vehicle for controlling and integrating the design effort. The main consequences of our assumptions about physical construction are that the model has to be able to deliver specifications for both the PCBs and the components on them. Of course, it also delivers the software for each processor in the system, but this is not the concern of this discussion.

Clearly, ICs can be bought off-the-shelf, or they can be custom built. The off-the-shelf (digital) ICs of interest to us will include: processors, memory components, communications controllers and collections of basic SSI/MSI components, such as gates and flip-flops. Analogue components come in similar packages and the implementation of some MOOSE objects will require a mixture of all these kinds of ICs. For example, a non-primitive object in a multimedia application might map on to a processor with memory, analogue amplifiers and signal modulators, and analogue-to-digital (and digital-to-analogue) converters. Custom built ICs can either be manufactured by a silicon foundry as Application Specific Integrated Circuits (ASICs), or they can be manufactured initially as general purpose programmable chips, such as Field Programmable Logic Arrays (FPLAs) to be customised by the system developers. An extreme example of an ASIC is where the complete computer system, hardware and software, is placed in a single chip; such an arrangement has become known as a Deeply Embedded System. Thus the final output of MOOSE has to support and accommodate all these variations in the final manufactured products.

So far, the transformational development path through MOOSE has been presented with mainly the ultimate degree of system level integration in mind; for example, by producing a single VHDL model for the complete hardware platform. In general, such VHDL models contain many 'entities', some specifying application-specific objects that are to be custom built, while others are only present to serve as emulators of previously engineered components. The latter might be factory provided macro-cells or bought-in ICs such as processors. If the VHDL is executed, the entities collectively provide an implementation-level simulation of the total system.

Entities in the VHDL model correspond to objects in the Platform Model, and when the Transformational Codesign is complete these should be labelled to identify groups that are to be placed in the same ICs. Some labelling of this kind is evident in the VCR model in Appendix 3. Within each group there may be a mixture of entities, some that have to be synthesised and others that represent the standard macro-cells. This pre-supposes that a foundry automation technique will manufacture chips that accommodate these mixtures. However, MOOSE is also intended to apply to much less sophisticated implementations in which the physical fragmentation of the implementation might be much greater and in practice may not involve any ASICs.

The 'lower-tech' implementations are necessary for several reasons, even though ASIC manufacturers are moving rapidly towards the goal of single chip solutions, automatically synthesised from VHDL specifications. First, despite impressive advances in the capabilities of silicon technology, for the foreseeable future there will always be some systems that are too large and complex to make the single chip solution feasible. Second, for some applications it makes good economic sense to use a mixture ASICs and standard ICs, such as processors and memories, rather than develop the fully integrated solution. A mixed solution of this kind might also be necessary for a system that contains analogue components, as in certain cases it may not be appropriate to mix analogue and digital objects on the same IC. Third, many systems, which are designed as integrated systems, need to have physically distributed implementations due to the nature of the application. Finally, the relatively high cost of developing ASICs may be out of proportion to the expected sales volume for a product, and an SSI/MSI implementation might be more cost effective.

It is thus envisaged that MOOSE models will be mapped to a variety of hardware implementations differing in sophistication and accommodating the range of needs stated above. These variations are derived by labelling the hard objects in the Platform Model with IC names. Obviously, if there is to be substantial fragmentation in an implementation, a large number of distinct IC labels will appear in the model, but its structure which reflects logical fragmentation does not change. In all cases except one, labelling objects with IC names is sufficient to specify the required physical partitioning of the design, and the number of different ICs that can be introduced in this way is limited only by the number of primitive objects in the model. The exception is where the implementation of a primitive object has to be distributed across more than one IC, which is the case when SSI/MSI components are to be used.

The approach recommended for dealing with such gate-level designs is to carry the design through to the final stages as for the single chip solution, then label the objects that are to be implemented at the gate level with the same IC name, a *virtual* IC. Further work is then required outside of the MOOSE paradigm, as described below, in order to provide a gate-level design for the virtual IC. An alternative, which is not precluded by any of the built-in features of MOOSE, would be to continue the hierarchical refinement of objects until primitives are reached that correspond directly to the logic components that are to be used. However, existing techniques and tools for design at the basic component level are well established, and the authors therefore recommend that the OID hierarchy is terminated long before the stage of introducing primitive objects such as gates and flip-flops.

If the recommendation for dealing with gate-level designs is accepted, the MOOSE model still plays its primary role of providing the functional specification from which implementation detail is produced. Also, at the point at which a conventional technique, such as logic schematics, takes over, the main benefits of the MOOSE approach will already have been obtained. A tested Behavioural Model will have been subjected to Transformational Codesign, and a Cosimulation environment will exist in which the software can be tested ahead of hardware development and the behavioural definitions of the main hardware partitions will have been tested.

9.3 IMPLEMENTATION OF HARDWARE

We will now summarise the actions necessary to complete a Platform Model to the stage where an implementation specification can be synthesised. First, the designer must think about the type of technology that is appropriate for each primitive object in the Platform Model, decide which ICs will be used and then define the packaging of objects into ICs by labelling each object with its IC name. Naturally the choice of the technology must satisfy the Design Constraints. It will often be the case that Design Decisions in the original specification leave little opportunity for choice, and consideration of Non-functional Requirements may also have led to firm decisions; for example, with regard to the use of specific processors.

The technology types that the designer must recognise when mapping the object model to ICs are as follows:

- An ASIC to be synthesised from VHDL.
- A foundry macro-cell to be placed on an ASIC.
- A standard 'bought-in' IC.
- A virtual IC (as defined above).

After a Platform Model has been suitably labelled, the MOOSEbench tools can synthesise two different outputs from a completed model, namely VHDL specifications and a net-list.

The VHDL output serves two purposes. First, it provides a hardware-level simulation of the complete system, which may be useful even if only a subset of the entities are to be manufactured from the VHDL, since the latter can be tested in a larger context. For those ICs that are to be manufactured rather than bought in, VHDL can be extracted selectively to provide a suitable input into either automated or manual processes. If fact the VHDL is only strictly necessary in the case where fully customised ASICs are required, and if a full simulation of the hardware is not required, effort might be saved during codesign by not completing the VHDL for components that are to be the bought in.

The use that is made of the VHDL output from MOOSE will depend upon the technology types that are to be used. For the extreme example of a fully integrated deeply embedded system, the VHDL will be input, almost unchanged, into the foundry development process for ASICs, ultimately to produce a single chip system. A less integrated implementation might require several ASICs, each implementing a part of the system, in which case the VHDL can be extracted in relevant sections, each of which can then be used to drive the manufacture of an ASIC either as a custom chip or as suitably programmed general purpose ASICs. If the implementation is to be developed by a manual process, the VHDL provides a detailed specification.

The second output from the MOOSE synthesis phase is a net-list, which defines the connections between the set of ICs named on the Platform Model. The net-list will be input into a conventional system for the manufacture of such PCBs as are required. Towards the extreme level of integration, where the implementation involves only a very small number of ICs, the net-list might be redundant. However, a typical large system would involve a the use of large number of ICs, using a mixture of technologies such as ASICs and standard bought-in ICs. In this case, the net-list provides a suitable input for an automatic routing package which will produce the artwork for the PCBs on which the ICs are to be mounted.

Of course, the net-list will also have to be transformed to introduce those components and connections that are concerned with providing power and decoupling in the system, but this *electronic* aspect of design is beyond the scope of this book. In hardware produced from VHDL, the synthesis tools can deal with these problems.

If the implementation is to be distributed over several PCBs, the net-list will have to be partitioned accordingly. This distribution of the net-list might require adjustments to each partition to effect the necessary interconnections; for example, by including sockets that connect to back-planes or wires. In fact, in many cases, the disturbance due to dealing with the interconnections between PCBs will be minimal since the ICs will be clustered around processors and inter-processor connections will have been developed explicitly in the Platform Model. In either case, the interconnection mechanisms will be standardised, for example as buses, and library objects are available that implement these standards.

At the opposite extreme to the single chip solution, an implementation may be required that uses only standard ICs, such as memories and processors, and the virtual ICs to be implemented at the gate level. In this case, the virtual IC entries in the net-list have to be replaced by other net-lists that specify their implementation in terms of the ICs which package the basic logic elements, such as gates and flip-flops. The net-lists will then be obtained from the SSI/MSI design tools used to develop the logic schematics that represent the implementation of the virtual ICs.

9.3.1 Producing Net-lists

The main problem to be dealt with in order to obtain a satisfactory net-list from a MOOSE model concerns the mapping of MOOSE information flows and event connections on to the pins of ICs. A subsidiary output from MOOSE assists in the solution of this problem. This output provides a list for each IC that specifies the connections to the IC in terms of their MOOSE names (the MOOSE connection list). In the case of information flows, their size in bits will have been specified in the CISs for objects in the Platform Model that receive and produce the flows, in which case a separate entry will exist in the MOOSE connection lists for each bit. Thus there will be almost a one-to-one correspondence between the entries in the MOOSE connection lists for an IC and its pins. The entries either give an event name or an information flow name with a suffix giving a bit number. For a net-list to be produced that uses pin numbers, it only remains for the developer to specify the equivalent pin number for each entry in the MOOSE connection list for each IC in the model. Until the pin assignment in the connection list is completed, the net-list that is produced will use the MOOSE name for any unspecified pin.

9.4 EVALUATING PERFORMANCE BY SIMULATION

Simulation in MOOSE is synonymous with executing a model. Chapter 6 described a simple implementation of an Executable Model based on a synthesised C++ program, which simulated the behaviour of a flattened network of primitive objects derived from an Executable Model. It made direct use of class definitions from the model to provide the primitive object behaviour and, as shown in Fig. 9.1, included three 'system' objects which are responsible for the information flow through the system, the external interface with the user and the management of concurrency within the model.

An implementation of the **Concurrency Manager** in the Executable Model was discussed which followed a policy of allowing individual threads of activity to run for relatively long periods, during which all other 'concurrent' activities waited, hence the accuracy of the simulation is open to some doubt. Another shortcoming of that method of simulation is that no account is taken of processing and propagation delays, and only a very approximate and arbitrary modelling of real time is provided. The time simulation was based on a count of the number of execution cycles, each involving different amounts of work and therefore corresponding to variable amounts of real time. However, despite its flaws, the simulation provides adequate facilities for validating logical behaviour and sequencing of actions, but not for evaluating actual timing and performance-related aspects of the model.

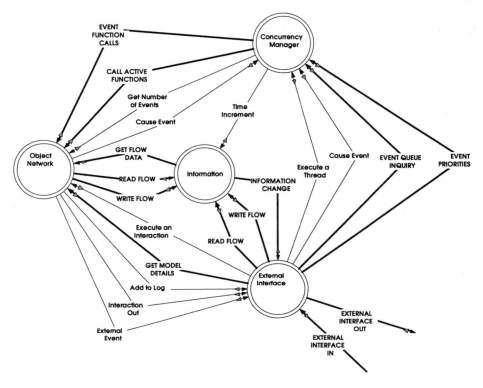

FIG. 9.1 The simulation system.

In fact no realistic simulation of real-time behaviour and performance could be provided by an uncommitted Executable Model since, by policy, none of the decisions that affect performance, such as the selection of hardware or software solutions, the type of hardware to be used, the number and type of processors and so on, have been taken at that stage. However, later in the development process when the commitment decisions are made, it is very important to know what effect they will have on performance. So that the impact of the design choices on critical aspects of timing and performance can be explored, a more sophisticated execution mechanism than the one described in Chapter 6 is presented here.

The most serious problem affecting the accuracy of a simulation arises in connection with modelling concurrency, and the options for dealing with this are constrained by the basic design of the execution mechanism. This hinges on the technique of compiling C++ class definitions built up from user-provided code for the functions of the application-specific objects, and executing this user-provided code under the control of the system objects that are automatically included in the simulation system in Fig. 9.1. A consequence of this technique is that, once the thread of execution is passed to a function in an application-specific object, no other activity can be started until control returns to a system object. However, from time to time the user functions will transfer control to the system objects, for example to access an information flow. These excursions through the system objects occur at the points at which an activity is likely to interact with others, and the pre-emption of the thread of control at these times gives significantly better results than the approach of waiting until the thread terminates by returning control to the **Concurrency Manager.**

An alternative approach to providing simulation would be to use an interpretive technique, in which case a new activity could be scheduled at any instruction boundary in the current activity. However, the compile and execute technique used in MOOSEbench maximises the speed of operation of the model, and replacing it by an interpretive mechanism could reduce the execution rate to a level at which execution of a full system model would no longer be viable. Improvements to the simulation mechanism introduced below retain the compile and run technique, and as a consequence the simulation of concurrency may still be approximate. Thus, to make safe use of a simulation experiment performed with the Executable Model, the user needs know more about the way it operates.

9.4.1 Timing Information in an Executable Model

The technique adopted to provide better timing information from an executing model requires the user to make timing estimates on a macro scale; that is, estimating the completion times of logically significant actions in a behavioural model. The simulation software determines their impact on timing at a system level by simulating the passing of time using the same coarse measure of time. Timing information returning from the system takes the form of 'time-stamps' appended to all outputs from the executing model, and these are in the same units that the approximations are given. To provide accurate simulation of timing at the micro level – that is, in units that relate to circuit switching speeds – would require implementation knowledge beyond the scope of MOOSE. It is expected that one use of the VHDL model produced by MOOSE would be to investigate these detailed timing issues with the aid of a VHDL toolset. MOOSEbench tools have a different goal consistent with their place in the lifecycle, which is to provide the means to develop confidence that a design, refined to the level of VHDL in the case of hardware, conforms to both the functional and non-functional requirements of a total system, before it is passed on to the hardware fabrication phase.

Thus the approach to timing relies upon active user participation, but as the user at this stage is a development engineer making critical code-sign decisions, it is reasonable to expect reliable input concerning the local impact of these decisions with respect to timing. For example, the user might provide estimates for the elapsed time between a stimulus arriving and a response being provided, between an event occurring and a related output change, or between two different output changes. This information is collectively called a *Timing Model* and it is placed mainly in the OSPECs of the primitive objects of the Committed Model. Obviously, the way in which this timing information is used within an executing model requires a number of issues to be resolved, and the user needs to be aware of this when planning and interpreting the results of experiments that use the model.

One issue, for instance, is, 'How should the given timing information affect the policy on pre-emption of threads of execution running through the model?'. For example, should a thread of execution be allowed to run forward in time beyond the current time and thereby produce post-dated output, or should it be suspended until the time step in which the output is due? Another issue is, 'Is it permissible for the threads executing sequentially to deliver post-dated outputs in other than ascending time order?'. Clearly, a descending order does not make sense in soft objects because the code producing the outputs in the simulation is the same as the code to be used in the implementation, and the timing of the outputs is determined by the code sequence and processing delays that result from execution of the code. In the case of hard objects it could be argued that, since the code in the executing model is a 'soft' emulation, it is a *sequence* of actions, whereas there is a possibility that the same actions might become *parallel* in the actual hardware implementation. Of more importance is the fact that, in an implementation, the operation of the hardware functions is very fast and the timing delays applying to the outputs are due to propagation delays rather than processing delays, which may again result in outputs becoming available in an order different to that in which they are produced. Therefore the threads of activity passing through hardware objects must be allowed to produce post-dated outputs and both ascending and descending time sequences have to be accommodated. Threads passing through software objects can be halted at the point where they attempt to produce post-dated output until the simulated time catches up. We would stress that halting a particular function within an object does not stop different threads entering the object.

To avoid getting too deeply involved in issues such as those outlined here, it is better to return to the level of how the simulation is controlled and how this affects the results. The discussion of underlying difficulties will therefore be cut short. Later in this chapter, the implications of the way in which the executable model simulates timing and concurrency will be considered.

9.4.2 Controlling the Simulation

For the purpose of conducting meaningful timing experiments with a Committed Model, it is necessary to understand how a simulation is controlled and, in effect, this means understanding the operation of the **Concurrency Manager**. This arises because there is no complete solution to the problems of simulating concurrency on a serial machine, and compromises have to made in the design of the **Concurrency Manager** and the other system objects that collaborate with it. In fact, in extreme cases, the user may need to customise the **Concurrency Manager** to deal with features particularly important to a given project.

We have already given some reasons for not treating hard and soft objects in the same manner. Another reason is that actual implementations of hard objects can be fully concurrent, whereas true concurrency in soft objects is limited by the number of processors available, and this is supplemented by the quasi-concurrency implemented by the Thread Managers of these processors. A final difference between soft and hard objects is that the former are inevitably subject to processing delays – that is, instruction sequences have non-negligible execution times – whereas hardware operates much more quickly but the output changes made in one object may be subject to propagation delays before they become available as inputs to another. These differences are taken into account by the way in which the **Concurrency Manager** models time.

The overall organisation of a simulation is that it operates in steps called execution cycles, each of which starts with the user of the simulation being invited to update all the inputs to the model. Of course, the user need not interact on every execution cycle, and provision is made to use pre-prepared scripts which supply the inputs, hence a simulation run need not be slowed down by user interaction time. If the user chooses to interact frequently the effect is that the simulation clock will apparently be stopped while the user responds, and time therefore passes more slowly, but the execution of the objects in the model stays in phase with the time simulation.

Assuming that no input values are post-dated, the information flow inputs from the user become current values for the cycle in which they are input, and event inputs are placed in queues inside the **Concurrency Manager**'s data space. We shall see later that these may be time-stamped with the time at which they are to be signalled and the queues are therefore ordered according to the time-stamp values, and they are streamed according to the events' priorities. To avoid going too far into detail, and because they rarely occur, we will ignore the possibility of interactions being input through the external interface.

After inputs have been updated, the purpose of the next step in an execution cycle is to execute the active functions and those event functions whose events have been signalled. Of course, if a previous activation of either has been halted to wait for the simulated time to catch up, new entries will be similarly delayed.

Typically, the operation of the **Concurrency Manager** treats each execution cycle as a constant interval in simulated real time, which is called *simulated time.* Simulated time is given a unit increment at the start of each execution cycle. All entries made to objects during an execution cycle are assumed to start at the time given by simulated time and, from the moment of entry, time moves forward but the **Concurrency Manager** must prevent it going beyond one unit of simulation time, otherwise interacting concurrent threads get out of step.

However, some operations that start at a particular simulated time may produce outputs which, for various implementation reasons, should not become available until some time later. For instance, an operation in a soft object can take a significant time to compute its outputs, depending on the speed of the processor and the amount of code that has to be executed. Hardware operations will take very much smaller times, but there may be a finite propagation time before their output becomes available to other objects. Thus all outputs from the objects in the model are time-stamped with a value representing the simulated time at which they are to become current. This time-stamping of outputs has obvious significance in the case of external outputs produced by an executing model, as it characterises the overall timing behaviour of the system. For example, from the time-stamps, time intervals can be computed between inputs and the outputs they stimulate or between two related outputs. Internally the significance lies in the fact that an output cannot become an input to other objects until simulated time has advanced to the value of the time-stamp. External inputs entering a simulation are normally delivered by the user at the appropriate simulated time, but they are also time-stamped, in order to maintain a consistent log of input/output activity.

A consequence of the behavioural differences discussed above is that the scheduling tactics of the **Concurrency Manager** must distinguish between hard and soft objects, as described below. First, we need to clarify how time information is managed inside an executing model

Managing the Simulated Time

Before a simulation is attempted, the user should provide a *Timing Model* for the particular set of implementation commitments being simulated. The information in this model specifies processing and propagation delays implicit in the commitment decisions. Specifically it defines the time at which each output from an object is to be expected, relative to a specified reference time. The reference time may be either: *start time*, which will be the time at which the thread of execution that produces the output enters an object; the quantity *current time*, which initially is the same as start time, but can be updated in several ways as the thread of execution moves on through the object; or it may be the time of the most recently produced value of any specified output or reference to any specified input. In fact, as a thread of execution moves through an object, it may produce more than one value on a particular output and hence a sequence of output times may be specified in the timing model.

There are three mechanisms that update current time, and their actions depends on whether the object using them is hard or soft. One mechanism which applies only to soft objects is that, when a time-stamped output is produced, current time advances to the time-stamp value. The rationale behind this is that outputs from software are instantaneous once they have been computed, but it may take time to compute them. Hardware objects will make significant use of a different facility, namely one which allows the specification of delays on outputs, but these will be assumed to be predominantly propagation delays and they are not coupled with processing times. The second mechanism is that a minimum elapsed time can be specified for each complete thread of execution through an object. This is mainly appropriate for stating a time for the active functions in soft objects. It is intended to facilitate the simulation of the sequential execution of active functions that is a necessary feature of the Thread Manager of a Platform Model.

The third mechanism is somewhat different in nature. Although it could apply to both types of objects, it is more likely to be relevant to soft objects and it provides a more flexible treatment of processing delays. It is based on a function call (in fact the built-in function **Time Call** mentioned but not described in Section 6.2.3), and it results in an interaction message being sent to the **Concurrency Manager** as shown in Fig. 9.1, which delivers a user-specified estimate of the time that has elapsed since the last change to current time. It is unavoidable that this mechanism is invasive, in the sense that it relies upon modifying the behavioural specification, whereas the other mechanisms are based on statements in the timing model contained in the OSPECs.

In summary, current time advances as a thread of execution passes through a model, either by a step change applied when an output is produced, or by the accumulation of time estimates based on the paths traversed; both methods of specifying the passage of time have their uses. It is important to note that the current time only advances inside any activity of the executing model as a result of entering or returning to a system object; for example, to produce output, access an input, or on completion of an active function. Therefore the system objects can decide whether the activity in progress should be suspended at the point of advancing its current time, in order to allow other concurrent activities to run and catch up.

Another feature of the Timing Model is that it allows an order to be specified in which the active functions are to be run. This provision for ordering active functions allows a user to specify a priority (and the minimum elapsed time estimate, mentioned above) for each active function of each object. These priorities determine the running order and, if no priorities are given, an arbitrary order will be chosen. The reason for ordering hardware objects is to allow functions that produce outputs with zero propagation delays to be placed first, since their outputs are to be available to other functions running at the same execution time. In the case of soft objects, the facility for ordering active function execution allows the effects of an order imposed by a Thread Manager to be simulated.

Scheduling Tactics of the Concurrency Manager

The purpose of scheduling is to organise an execution cycle during which the appropriate functions are called in objects that have work waiting to be done. A number of issues arise concerning the identification of these objects but, in principle, all the active functions should be given the opportunity to decide for themselves, and event functions should be called if their events have been signalled. Also, the time that functions need to complete and the order in which they are called have to be considered, since they are supposed to run concurrently. The description becomes more simple if we focus first on the treatment of their active functions, even though, in practice, priorities might dictate that the event functions are to run first or be intermingled with the active functions as described below. Two categories of active functions, hard and soft (which, for this purpose, includes firm), are distinguished.

The hard functions are run once per execution cycle, sequentially and in priority order. This tactic models the behaviour of the hard objects of an implemented system quite well, in that they typically operate concurrently and very quickly, even though their outputs may be subject to propagation delays. If post-dated output is produced, it is queued until the corresponding simulated time, but the active functions are allowed to operate again in the meantime, possibly adding other outputs to the queues.

We have already seen that the soft objects in an implemented system are assigned to processors and each processor has its own Thread Manager, which determines the sequence in which its active functions are run and how event priorities are handled. The treatment of soft functions during a simulation should model this. Thus, soft functions are treated in groups according to the processor on which they run. Each group is run sequentially in priority order but, in any one cycle, the sequence only proceeds until the current time is moved more than one increment ahead of simulated time. At this point, the function is suspended until simulated time catches up. Completion of the full cycle of active functions can take a significant time, which is modelled in the simulation by allowing it to spread over an appropriate number of simulated time intervals.

There are three points at which the execution of the group of active functions assigned to a processor can be suspended to wait for simulated time to catch up. First, on exit from an active function and after enforcing the lower bound on its execution time, if current time is ahead of simulated time by more than one increment, the execution of the next active function is delayed until the execution cycle in which simulated time catches up. Other points at which the **Concurrency Manager** could regain control and hence invoke a suspension are when the **Time Call** function is entered, and when an output is produced or an input is read. In all three cases, the thread of execution will be in a nested sequence of function calls. The **Concurrency Manager** therefore has to use techniques similar to those of the task changing mechanism in a kernel to preserve the current context. The conditions that cause this action are either that current time is moving more than one increment ahead of simulated time, or it is one increment ahead and an input with an earlier time-stamp is about to be read.

The occurrence of events produces no significant new problems for the scheduling tactics, but the user should know how they are handled. Events like active functions are given priorities. When an event becomes effective – that is, simulated time is equal to its time-stamp – it is a candidate for being scheduled. If its priority is greater than the current activity, it is allowed to pre-empt the current activity; otherwise it remains queued. The queues of events are inspected whenever the **Concurrency Manager** is making a decision about which operation of an application-specific object should be called next. We should also remember that events may be serviced by more than one object. Event functions that extend their current time beyond the simulated time of the next execution cycle will be suspended in the same way that active functions are suspended.

9.4.3 Potential Problems in Timing and Concurrency Simulation

There are potential problems in correctly simulating the concurrency of a model wherever there is information transfer between concurrent threads, and a timing model is defined that uses time increments that exceed the propagation time of some outputs. In effect this means that some outputs have zero propagation times, when measured in the (integer) time increments of the simulation. Clearly, reducing the time increment would eliminate the problem, but it slows down the simulation and, since the simulation is modelling a complete system, there is conflicting pressure to make the time steps as coarse as possible. A solution that might avoid the need to suffer the consequences of a very short time step is to make use of the priority mechanisms in the timing model to force an execution order on the threads of execution of the active and event functions that matches the direction of information transfer.

The problem arises in hard objects that have active functions monitoring information flows for the purpose of acting on significant changes. Since active functions are supposed to be running concurrently it is reasonable to expect a change made by one to be soon noticed by another. However, they may be run in an order that allows such a change to go unnoticed until the next time step. Obviously, if the change is subject to a non-zero delay, the modelled behaviour will be as expected. If there is a zero propagation delay the reverse order is required. Clearly, this solution is not possible if there are loops in the information flows or bidirectional flows between objects.

In the case of soft objects an analogous problem can arise with information flows between objects running concurrently in separate processors. These information transfers should normally be controlled by some form of explicit synchronisation protocol, particularly if the successful operation of the system depends on the synchronisation of producer-consumer relationships between software running in independent processors. The problem does not arise with soft objects allocated to the same processor, since the active functions run sequentially in both the implementation and the simulation, hence it is sufficient to force the same order in both

Although errors in a simulation of the kind we have discussed cannot be corrected automatically, they can be detected. The detection is based on keeping a record of the latest time of reference to an 'old' information flow value. When such a value is updated from new output, if the last reference is not earlier than the new time-stamp then an error has occurred, or at least an out-of-date value of the information flow has been used.

Another potential problem concerns making proper allowance for the effect of losing time in a soft function due to the processor being preempted. The safeguard against this is that, when activities are interrupted by events, their current time stands still until they are resumed. Thus each object experiences an unbroken flow of simulated real time.

The conclusion to this discussion therefore is as stated earlier: that, in order to make safe use of the MOOSE type of full system simulation, users must be conscious of the departures between true system timing behaviour and the way in which an executing model approximates to it, and a realistic timing model has to be specified that makes proper allowance for propagation delays in hardware and processing delays in software.

9.4.4 Using Simulation to Support Codesign

It is worth reminding ourselves at this point of the implications regarding the completion of the Codesign Phase on the status of the development. With respect to software, the implication is that complete source code can be produced for each processor in the system which has been tested in a simulation of the full system provided by execution of the Committed Model. The assumption about hardware is that a net-list can be produced automatically, which can be used to drive the production of any hardware that is required in the form of PCBs. Additionally, VHDL can be produced that will support a simulation of the operation of the complete hardware platform and also electrical level simulation and synthesis for any ASICs that have to be manufactured.

Clearly, the system simulation experiments conducted with the Committed Model have to give a high degree of confidence that both the software source code and the VHDL are correct and complete from the point of view of both functional and non-functional requirements. Functional issues should mainly have been settled before the codesign phase was entered and, during codesign, the experiments should focus on exploring performance issues, checking that the timing constraints can be met and verifying that the design and commitments are sound from a timing point of view.

10 Concluding Remarks

The objective of the authors has been to explore the issues that affect the successful development of computer systems-based products and to demonstrate an integrated approach to engineering the computer systems contained in them. We have had in mind any product that makes use of special purpose computer systems, or more generally microelectronic technology, and physical implementations ranging from single chip (deeply embedded) solutions to widely distributed systems involving multiple processors closely integrated with application specific hardware and software components.

Integration in the approach has been sought through models that are refined incrementally until they become specifications in standard implementation languages. All stages of the process, from the capture of initial ideas through to entry into typical manufacturing tools such as compilers and simulators for C++ and VHDL, are covered. Quality control is imposed through stage-by-stage validation, which monitors the passage of a single model through a prescribed sequence of transformations leading to the final goal of automatic manufacture from verified source code.

The concluding remarks in this chapter summarise the status of the MOOSE paradigm, the notations and tools that support it and the authors' experiences in applying them.

Central to the approach is a notation that allows systems to be described as hierarchically clustered collections of objects, each of which can be implemented in either software or hardware, except perhaps on the analogue fringe where choice may be limited. The notation has developed through experience in its application. After a short period of evolution, during which appropriate connection semantics for such objects were discovered, the notation has stabilised and the authors are now unaware of any behavioural requirements that cannot be modelled in a style that is open to optimum implementation. The workbench provides for the capture of these behavioural models in a similar manner to that of well known CASE tools.

As in most other model-based approaches, experience in applying the notation to industrial designs has indicated unbounded scope for exploiting human skill and experience while constructing the models, but a promising start has been made on distilling this experience into guidelines that make model building almost deterministic. Obviously the quality of a product whose architecture follows that of a model is heavily dependent upon the right structural decisions having been made in the model, hence there is much at stake in the reviews that form an integral part of the approach. During such reviews, the hierarchical structure of models is very important, as is the abstraction of interconnection detail, since it is these features that make the models readable. However, readability alone is not enough and, in the authors' experience, the mechanism provided for simulating dynamic behaviour by executing the models is even more essential.

The precision in detail that an executable model must necessarily contain offers other opportunities that have not yet been exploited by the MOOSE workbench, particularly concerning static evaluation of structure and static analysis of both performance and behaviour. These automated operations are the subject of research that will lead to future extensions to the facilities of the workbench.

Another big issue that influences the success of approaches such as the one we have described is their cost effectiveness. Research projects sponsored by both EPSRC and ESPRIT OMI are in progress to obtain more evidence and experience on this issue. Current experience is limited to about ten models, most of which were constructed while the notation was evolving. Clearly, it would be inappropriate to attempt to distil to much out of this limited experience except for the following, which has become very clear:

- Making good models is difficult.
- Making systems without models is at least as difficult.
- Productivity in model making increases with experience.
- Apart from the need to iterate with models, none of the work is 'throw away'.
- Previously modelled systems are a rich source of re-useable designs and implementations.
- Reusable component libraries dramatically reduce the effort and timescale of new projects.

In the long term the success of approaches such as MOOSE must depend upon their success in spawning re-useable designs/implementations from the occasional pioneering projects that break new ground. For example, the most cost-effective products will be those that can be fabricated from parts of simpler products or developed as product ranges from a central design model. The nature of the various stages of modelling in MOOSE lends itself to integration of a pragmatic library structure with the MOOSE workbench, in which behavioural models can be provided to assist the assimilation of library items into new designs without releasing proprietary implementation source code.

APPENDIX 1

MOOSE Workbench User Guide

A1.1 INSTALLATION AND OPERATION

MOOSEbench runs under Microsoft Windows, version 3.1 or above, on any IBM compatible personal computer with the following (or better) specification: 486-DX, with at least 1 Mb of free hard disk space, 4 Mb of RAM, VGA graphics (640 x 480, 16 colours) and a mouse or equivalent pointing device that operates through Microsoft Windows. It will be assumed throughout this guide that the user is familiar with Windows and only the MOOSE specific features of MOOSEbench will be described.

A1.1.1 Installation

A file, **moose.exe,** containing MOOSEbench, is available through the Web page **http://www.cl.co.umist.ac.uk/moose**. MOOSEbench can be installed by obtaining a copy of this file, and creating a 'Program Item' entry for it using the normal procedure for adding new applications to Windows, after which a MOOSE icon should appear and MOOSEbench can be run by clicking on the icon in the normal manner.

The first time the tool is used an empty directory of projects will be created in the file 'moose.ini'. Projects may then be added, and will be available for editing when the tool is used subsequently. A sample project is also available through the Web. This is a complete 'Executable' Model of a VCR, and is stored in file vcr.mwp. Users who wish to experiment with the workbench may prefer to install this model on their machine using the Retrieve option described in Section 2.3 rather than creating their own project from scratch.

A1.1.2 Using MOOSEbench

When MOOSEbench is started the user is presented with the main window shown in Figure A1.1, which displays a list of existing projects (the

Project List). This window has a standard Microsoft Windows format with the usual controls for manipulating it, and it can be closed by using the system menu button.

Selecting the *File* entry in the menu bar reveals a pull down menu with the choices *New* and *Exit*, the latter being an alternative to the system menu button. The operation of *New* is described below.

The *Project* menu offers the options: *Capture, Synthesise, Rename, Archive, Retrieve Delete* and *PrintMD*. All operate on the project that has been **highlighted,** by selecting it in the **Project List**. The last five operations deal with complete projects as described in Section 2. Sections 3 and 4 discuss capture mechanisms, which covers the input, editing and browsing of MOOSE models. Section 5 defines the form of various textual specifications that are necessary to support the synthesis operations, which are summarised in Section 1.2.2.

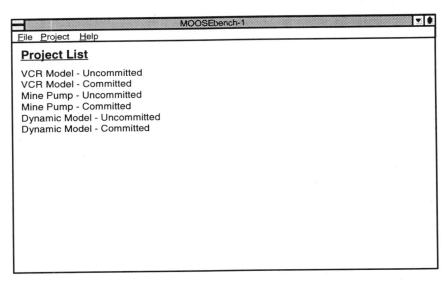

FIG. A1.1 The MOOSEbench main window.

A1.1.2.1 Creating a New Project

A new project is created by selecting the *New* option, which displays a dialogue box requesting the user to enter a title for the project and a 'DOS' pathname for the file where its detail will be stored. The title may be up to 40 characters long, and if it is already in use it will be rejected. A full DOS path must be given (including a file name) e.g. c:\mooseprj\vcr, and warnings are issued if there are errors in the path name, or if a file of the same name already exists.

When the new project name has been accepted, it will appear in the *Project List*, but there will not be a corresponding file until either the capture option is used or an archived project is retrieved as described in Section 2.

A1.2.2 The Synthesis Operations

In a full MOOSE workbench source code to implement both executable models of a system and the system itself can be synthesised from a model whose detail has been sufficiently developed. This version of the workbench offers only one such operation, namely the synthesis of a C++ program to implement the Executable Model. Selection of this operation produces a dialogue box requesting a path name and file name for the required source code. When the information is given a background task is initiated to generate the file, and the progress of this task is monitored. The use of these source files is external to MOOSEbench, for example it is within a standard C++ development environment.

A1.2.3 The Help Facility

Help facilities that conform to Windows practise may be selected from the main menu bar. Specific topics may be accessed in the usual manner.

A1.2 OPERATIONS FOR MANIPULATING PROJECTS

These are the operations *Rename, Archive, Retrieve* and *Delete* in the *Project* menu and they operate on the project which has been **highlighted** in the *Project List*.

A1.2.1 Renaming a Project

This operation assigns a new name to an existing project via a dialogue box. If the new name is already in use a warning will be issued.

A1.2.2 Archiving a Project

This operation copies the file containing a project's detail to any specified file in the file store or on a private disc. The project and its entry in the *Project List* will remain unaltered and no record is kept of the archived copy, although it can easily be identified since it will retain its '.mwp' extension. Thus a user may take any number of backup copies of a project for security reasons or at significant landmark positions in the development of a project, for example at the start of the Codesign phase. The

Archive dialogue box that appears when the operation is selected will contain the name of the project to be archived, and it requests a full path name for the file that is to become the archive copy.

A1.2.3 Retrieving a Project

Retrieval of an archived project is achieved by assigning it a name in the *Project List* using the *New* operation, and selecting *Retrieve* whilst this name is **highlighted**. The user will be prompted for the full path name of the archive copy. If it is on a floppy disc, the disc must be in the disk drive. An error is reported if the file cannot be found, and the user is warned if the copy is to a project that already exists.

A1.2.4 Deleting a Project

The *Delete* operation removes the **highlighted** name from the *Project List* and deletes the associated project file, but first the user is asked to confirm the request.

A1.2.5 Printing the Model Dictionary

Sometimes the information in the *Model Dictionary* may be required in hardcopy and the *PrintMD* operation provides this.

A1.3 ENTERING THE CAPTURE FACILITIES

Selection of the *Capture* operation from the *Project* menu opens the highlighted project to display its components. Alternatively, the project can be entered by double clicking on its entry in the *Project List*. Once a project is open, components may be added, deleted or modified. An *Item* option is added to the main menu bar and the window displays the *Project Directory* of the selected project as shown in Figure A1.2, which is the Project Directory for the VCR example mentioned in Section 1.1. Selection of the *File* menu will reveal that a *Save* operation has been added and there will be a *Close* option in the *Project* menu.

A *Project Directory* serves as both a 'contents list' for a project and an entry route to individual items, namely the *Functional View*, the *External View*, all the OIDs and OSPECs, and via the *Class Index* to class definitions. The OIDs are ordered and indented according to their position in the hierarchy, and the entries for the primitive objects are marked ('*') to make them easily recognisable. Both may of course also be accessed by

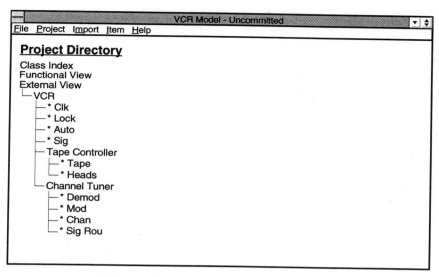

FIG. A1.2 Window after opening the VCR project.

navigating through the diagram hierarchy from any higher level OID, for instance the *External View.*

Only the first three entries appear in the *Project Directory* of a new project, and these will be associated with null items until they are edited. OID entries are automatically made as the OIDs are created, by the mechanism described in Section 4.1.

The *Class Index* provides an alphabetically ordered list of the class definitions, and new entries are added as described in section 3.1. The number of class definitions is not limited to those used in any specific version of the model, and there may be different versions of a class definition to support different simulation experiments, but all class names must be unique. The operative definitions at any time will be those whose names are referenced from the OIDs (and possibly CIDs if inheritance is used). Entries in the *Class Index* are described in the next section.

A1.3.1 Operating on Items Within a Model

The *Item Menu* provides operations *Edit, Rename, Delete, Print, New Class and Copy Class* and the first three can be applied to any item in a model that can be selected from the *Project Directory* or *Class Index,* whereas the last two apply specifically to classes.

Selecting an Item

The *Functional View, External View,* any OID or OSPEC can be selected

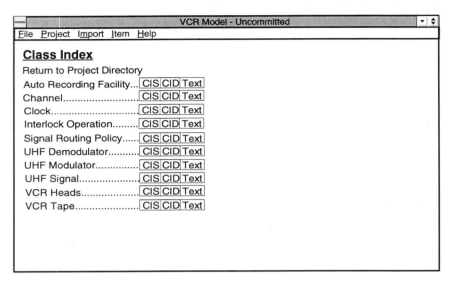

FIG. A1.3 The project class directory.

by choosing their entry in the **Project Directory.** Selection of the **Class Index** entry will make the components of class definitions available as shown in Figure A1.3, which relates to the VCR example. A return to the **Project Directory** can be made by selecting its entry at the top of the **Class Index.**

Class definitions have a one line entry containing the class name and CIS and CID fields. Either the complete entry or the CIS or CID fields can be selected by clicking at the appropriate positions.

Editing an Item

Editable items are those that can be selected directly from the **Project Directory** and the CIS and CID components in the **Class Index.** When the **Edit** operation is chosen, the appropriate Editor for the item is entered. In the case of diagrams this is the MOOSE Diagram Editor described in Section 4. The textual items, for example OSPECs and CISs, are placed in dialogue boxes to be edited with standard Windows facilities.

Renaming an Item

Any selected item can be renamed and consequential changes will be made throughout the model. For example, if an OID is renamed the reference to it on its 'parent' diagram will reflect the change. When the Rename operation is chosen a dialogue box will appear showing the old name and requesting a new one.

Deleting an Item

After the usual request for confirmation has been sought the item will be deleted, and so will any descendants. For example, if a class is deleted this removes the CIS, the CID and the FSPECs and DSPECs associated with the CID; similarly if an OID is deleted all the hierarchy below it including OSPECs is removed. Thus deletion is a drastic step and selective change using the editor is more common.

Printing an Item

A hardcopy of any item in the **Project List** may be printed using the *Print* operation, this includes the *Functional View, External View,* any OID or OSPEC and any CIS or CID. Apart from the normal windows printer set-up options that are provided, in the *Print Dialogue Box* that appears, the user is also given the choice of printing a single item or the item and the hierarchy of documents associated with it. For example, in the case of an OID this would include the OID and all the diagrams below it, up to and including the OSPECs and in the case of a CID, the diagram itself and the definitions of all its functions and data stores would be printed.

Introducing a New Class Name

Class names are introduced by the operation *New Class* which provides a dialogue box requesting a name, which must be unique. This name is entered in the **Class Index** and null items are created for its CIS and CID, which can then be defined using *Edit.*

Copying a Class

A common requirement is to have variants of a class definition, and these may be of temporary interest, for example in connection with a series of simulation experiments. The *Copy Class* operation provides a convenient way of generating these variants. Of course, if the classes are of permanent interest it is better to make use of inheritance to introduce the common aspects. Depending on how the item is **highlighted**, the copy can apply to the CIS or the CID or to both of these. A Dialogue box requests the name of the destination class, which should have been created with the *New Class* option and so exist in the **Class Index**. Note that the destination class should be undefined prior to the copying operation.

A1.3.2 Saving an Open Project

In order to avoid losing changes that have been made to an open project, they can be saved on demand by using the *Save* operation in the *File Menu*. Since saved changes cannot be 'undone' users should periodically make archive copies of a project under development, in case it becomes necessary to reverse changes.

A1.3.3 Closing a Project

The *Close* operation in the *File Menu* will produce a dialogue box to ask if the outstanding changes are to be saved. After the choice has been made, the project is closed and the *Main Menu* and **Project List** re-appear.

A1.4 THE MOOSE DIAGRAM EDITOR

The MOOSE Diagram Editor is entered if the *Edit* operation is selected from the *Item Menu*, with a graphical item of the project **highlighted** (double clicking on the item has the same effect). On entry the window will contain the diagram for the selected item and the menu bar will provide the options *File, Project, Item, Node, Connection, View* and *Help,* as shown in Figure A1.4. Options in the *File Menu* will be *Save* and *Exit,* and the *Project Menu* will contain a *Close* operation. The *Item Menu* will offer *Print, Ascend* and *Exit* options, the former referring to the item rather than the whole project. *Ascend* is concerned with moving upwards in a diagram hierarchy and it complements a *Descend* operation in the *Node Menu*. Both operations are described below. The *Exit* operation leaves the editor and returns to the 'open project' state with the window displaying the **Project Directory.**

All MOOSE diagrams consist of nodes connected by lines, and the *Node* and *Connection* menus allow these two features to be manipulated. The details of their operations depend upon the type of diagram selected but, since the basic operations are similar, a general description is given in Sections 4.1. Points relating to specific diagram types are presented in Section 4.2. Some general terminology is summarised below.

A **node** is a symbol on a diagram to which lines can be connected. Examples of nodes are the symbols representing states on an STD; objects on an OID; functions and inherited classes on a CID and externals on an External View.

Connections are lines that connect nodes. Different types of diagram require different types of connections, drawn in a variety of styles. However, lines usually have a direction, a starting node and a finishing node. They may also be composed of segments passing through way points.

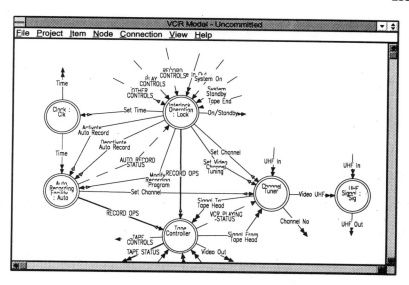

FIG. A1.4 MOOSE editor window.

External connections are lines that are only connected to a node at either the destination or source end, the connection of the other end being shown on the parent diagram.

Associated with lines are textual **labels** that have definitions in the **Model Dictionary,** the style of which will depend upon the type of the line.

Any area in a diagram devoid of other symbols or lines is called the *Background.*

A1.4.1 Creating and Editing Diagrams

MOOSE diagrams are created automatically when they are edited for the first time. The *External View* and *Functional View* are therefore created when their entries in the **Project Directory** are selected and *Edit* is chosen from the *Item Menu.* OIDs in general are created by applying the *Descend* operation to an undefined object node, as described later. CIDs are created by applying the *Edit* operation to the CID field of an entry in the *Class Index.*

The initial form of a diagram is determined by its type, which is implicit in the entry path used, and the basic principle is that the editor places on the diagram what is already known about the item. For example, a *Functional View* has at least one state and so it is created with one state, an *External View* has only one object hence this is provided, an OID is given the external connections corresponding to the lines connected to the symbol representing it on a parent diagram, and a CID has external connections as defined in its CIS. Section 4.2 discusses the treatment of external connections.

Diagrams are developed using the editing operations contained in the *Node* and *Connection* menus. This section discusses the complete set of operations that appear in these menus, although not all these options apply to each diagram type, for example the *Commit* operation only applies to OIDs. Since in practise the same operations will often be used repeatedly, first the required operation is selected and then it is applied to any nodes or lines which are subsequently picked. For example, if it is required to delete some nodes in order to revise a diagram, the *Delete* operation should be selected and then any number of nodes can be deleted by picking.

A1.4.1.1 Editing Nodes

The operations available in the *Node Menu* are:

- *Create* for creating a new node.
- *Rename* for changing the name of a node.
- *Connect* for making a connection to a node.
- *Move* for moving a node.
- *Delete* for deleting a node.
- *Commit* for committing a node.
- *Descend* for descending the hierarchy.

To Create a New Node

First select the *Create* operation, then click on a position in the *background* to indicate where the node is to be placed. A dialogue box will appear requesting a name and type for the new node. The choice of type offered will depend upon the type of the diagram. For example in the case of an *External View* only the type External is allowed since the only other permitted node, an object, will have been placed on the diagram when it was initialised. After the type has been chosen and a name given, the node will be drawn at the indicated position.

Renaming a Node

A node may be renamed by first selecting the *Rename* option and then clicking on the node. The dialogue box that appears depends on the node type. For example, if the node represents a non-primitive object or a function the user is asked for a new name, if it is a primitive object the user will have the option of changing both the object name and the class name.

Adding a Connection Between Nodes

With the *Connection* operation in the *Node Menu* selected pick the node at which the connection is to start, holding the mouse button down. This will produce a line from the selected node to the cursor position. Drag the free end of the connection to the destination node and release the button. A dialogue box will appear requesting a label for the line, a comment on its purpose and a line type. Once the required information has been provided the connection will appear as a straight line drawn between the two nodes. If it is desirable to route the line by a more indirect path, waypoints may be added, as described later.

Moving a Node

To reposition an object node select the *Move* operation, pick the node holding down the mouse button, and drag the object to its new position before releasing the button. Any line connected to the node will move with it. The free end of an external connection may be repositioned in the same way, as if it were connected to an invisible node.

Deleting a Node

To delete a node, first select the *Delete* operation and then pick the node to be deleted. Confirmation will be requested before the editor proceeds with the deletion. If the deleted node has connections they will remain and the ends previously connected to the deleted node will be free ends that can be connected to other nodes or the connections can be deleted as appropriate.

Committing a Node

The only nodes that can be committed are the objects on OIDs. These can be given a particular commitment status by selecting the *Commit* operation and choosing the node representing the object. A dialogue box will offer the four commitment choices: *uncommitted, soft, hard* and *firm*. If the object whose commitment is changed has descendants whose commitment status is not compatible with the choice, a warning will be produced and the option will be offered of having them made compatible. Compatible means that either all the descendants of an uncommitted object are uncommitted or, it has at least descendants committed in different ways. All the descendants must be committed the same way. If all the children of a parent object are committed to the same type the parent will be automatically committed to this type.

A1.4.1.2 Editing Connections

The following operations for editing the connections on diagrams are available in the *Connection Menu* :

- *Reconnect* for changing connections.
- *Direction* for changing the direction of a connection
- *Waypoint* for creating waypoints.
- *Cancel Waypoints* for removing waypoints.
- *Relabel* for change a label.
- *Define* for defining the detail of a connection.
- *Unbundle* for substituting the components of a bundle.
- *Delete* for removing a connection.
- *Tag* for implementing a connection via tags.
- *Cancel Tags* for replacing tags by a line connection.

Moving Connections

A connection to a node can be detached by selecting the *Reconnect* operation then picking a line near to its point of connection. By holding the mouse button down, the connection can be dragged to its new position and the button released.

Changing Direction

The direction of a connection can be reversed by selecting the *Direction* operation and then picking the connection.

Adding Waypoints

By using the *Waypoint* operation a waypoint can be placed at any point on a connection by selecting the connection and holding the button down whilst dragging the waypoint to the required position. Repeated use may be required (up to a maximum of 10 waypoints per connection) to establish a satisfactory route for a connection.

Removing Waypoints

Waypoints on a connection can be selectively deleted by selecting the *Cancel Waypoints* operation and then picking waypoints that are to be removed.

Labelling Connections

A connection may be labelled by first selecting the *Relabel* operation and then clicking on the connection. The dialogue box that appears allows both the label and comments associated with the connection to be changed.

Defining Model Dictionary Entries for Connections

The purpose of the *Define* option is to define Model Dictionary entries for the connections on OIDs which may be: *events interactions, parameterised events, time continuous information flows, time discrete information flows,* or *bundles* of any of the above, including *bundles* of *bundles* or *mixed bundles.*

All connections except *events* may have a structure that requires definition. For example *interactions* and *parameterised events* can have parameters, *information flows* can have fields and *bundles* have constituents. We shall call these the components of the connection.

To make a model dictionary entry first select the *Define* operation and then pick the connection. A *Dictionary Definition* dialogue box specific to the connection type will then appear headed by its name and type followed by a formatted definition, a component list and finally buttons that allow components to be edited (*Edit*) or deleted (*Delete*).

Accessing Components of Bundles

On any diagram, access to the components of the bundle can be obtained by using the *Unbundle* operation. This can be applied by clicking on the bundle, which is then replaced by a separate line for each of its components.

Deleting Connections

To remove a connection, first select the *Delete* operation and then pick the connection to be deleted. The user is then asked to confirm the request before the deletion is made. If the connection has been propagated to a lower level diagrams (as an external connection), this also be deleted.

Connecting Through Tags

It is sometimes difficult to make a connection between nodes on a diagram without causing crossovers or obscuring detail. To avoid this a break may be made in the connection by selecting the *Tag* option and then clicking on the connection. The connection will then be broken and an identifying tag placed at the broken ends. These ends may then be placed in a convenient position in the diagram using the *Move* operation described above.

Removing Tags

To remove tags from a connection, first select the *Cancel Tags* option and then pick either half of the broken connection. The tags will be removed and the connection drawn in full.

A1.4.1.3 Moving around a Diagram Hierarchy

Several types of MOOSE diagrams can have a hierarchical structure. Furthermore the diagram hierarchies end at primitives that have textual definitions. For example, objects on OIDs are defined by other OIDs and eventually by textual OSPECs. Functions on CIDs are also defined by textual FSPECs and data stores by DSPECs, and states on STDs might be defined by other STDs or textual SSPECs. The *Descend* operation in the *Node Menu* allows a user to move down such hierarchies, and also to create them.

To move to the diagram or text defining a node select the *Descend* operation and pick the node. The result depends upon whether the node is defined as a diagram, as text, or is undefined. In the first case the current diagram in the editing window is replaced by the new one. The effect in the second case is to present the text in a dialogue box in the same way that OSPECs entered from the **Project Directory** are presented. In the case where no definition exists and either text or diagram is permitted a dialogue box will appear inviting the user to specify which is required. This applies for example to an object on an OID which may be primitive or non-primitive, hence it may be defined by either an OSPEC or an OID. A side effect of choosing text in this case makes the object primitive, and the name on the object will then be preceded by a space for a class name terminated in a ':' .

After moving down a hierarchy by means of the *Descend* operation a return to a previous level can be accomplished by selecting the *Ascend* operation in the *Item Menu*.

A1.4.1.4 Moving Around Diagrams

Scrolling

To access parts of the diagram that are outside the current view of the screen use the Windows scroll bars surrounding the screen.

Extending Boundaries

When the boundary of a diagram has been reached through use of the scroll bars, and more space is required, further clicking on the scroll bar will bring up a dialogue box that gives the user the option of extending the diagram boundary.

Zooming

There are several options available from the *View Menu* for resizing a diagram in the editor window. These are *Auto Zoom*, which reduces a diagram to a size that will fit completely into the editor window, *Zoom In* and *Zoom Out* which allow the user to magnify or reduce the part of a diagram centred in the editing window, and finally *Set Zoom* which allows the user to mark the required zoom area by surrounding it with a box. The box is defined by first clicking at the desired point in the diagram for one corner of the box and then dragging the pointer to the position of the diagonally opposing corner and clicking on that point.

A1.4.2 Points Concerning Specific Types of Diagrams

The general description of the Editor given above has omitted some of the detail that applies to specific kinds of diagram. For example, the choices offered regarding node types will be those that apply to the type of diagram being constructed. The four types of diagram are recognised by the tool, and the allowable node types for each diagram are:

Functional View, permitted nodes: *state, state machine.*

External View, permitted nodes: *external object, object.*

OID, permitted nodes: object, *library object.*

CID, permitted nodes: *function, control function, data store, inherited class.*

The position with respect to connections is less clear cut, particularly with external connections. All diagrams have internal connections consistent with their type, and the external connections brought into a diagram from the parent level must be consistent with those on the parent diagram. However, the types of connections permitted on the parent diagram may be different from those allowed on a child diagram. For example, an OID may be the child of the *External View*, and although time discrete information flows are allowed on the latter, they are not allowed between objects on an OID. Thus different rules may apply to the connections on an OID depending on whether they are internal or external, and the editor will enforce these rules. Similar considerations apply to a CID whose external connections will be those specified in the CIS, but the internal connections are limited to function calls and data access. The total set of logically different connections to be accommodated are: *transitions, events, parameterised events, interactions, information flows (time discrete/continuous), data accesses, function calls* and *bundles.*

A1.4.2.1 Defining the Structure of Connections on OIDs

Section 4.1 described the creation and manipulation of connections on OIDs and briefly introduced the facilities for defining *Model Dictionary Entries* for connections. The following section describes in detail the use of the *Dictionary Definition Dialogue Box* to define the components of connections, by manipulating the multi-line field, referred to as the *Component List*, in combination with the *Edit* and *Delete* buttons shown on Figure A1.5.

Adding Components

A component is added by selecting the first available blank line in the Component List and then clicking on the *Edit* button. In the box that appears either a *textual description*, enclosed in quotes, or a name, may be entered. *Textural descriptions* are used to provide a string specifying the component, that used when CISs are defined, to help make decisions about a connections type and structure. This option is therefore not available when the connection type is a bundle and the component may only be another connection. A *name* may also be used to represent the information structure of a parameter or result.

As entries are made they will appear individually in the *Component List* and a collectively in a *Formatted Definition* line, which uses a format appropriate to the connection type.

Defining Components

Each entry in the *Components List* may be defined or deleted by selecting it and then clicking on the *Edit* or *Delete* buttons respectively.

When defining a string the user is presented with a box for editing the string. When defining a *name* another *Dictionary Definition* dialogue box will appear, again specific to the type of the component the name represents. If this is an information name the box will be the same as that used for *information flows*, allowing the user to define the fields of the structure. If the name is a constituent of a single type bundle the new dialogue box will automatically be headed by the bundle's type. Alternatively if the bundle is of mixed type an intermediate dialogue box will appear, for selecting the connections type, before the *dictionary definition* dialogue box is displayed.

Thus all the components of structures or bundles may be progressively defined through the above process. Alternatively the individual connections in bundles may be defined by picking a connection on a lower diagram in which the bundle has been decomposed.

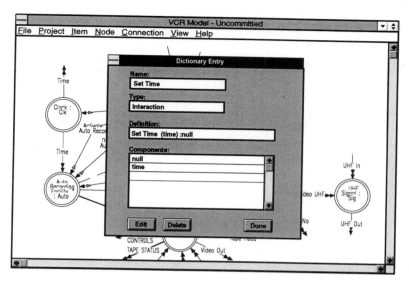

FIG. A1.5 Defining a connection.

A1.4.2.2 Positioning External Connections

These are relevant to OIDs and CIDs. For example the connections entering or leaving an object on a parent diagram are automatically placed on
its OID diagram. Incoming connections appear on the top left hand side
of the diagram and outgoing connections on the top right as shown on
Figure A1.6 which is the initial version of the OID defining the Channel
Tuner in Figure A1.4.

To attach incoming connections to an object on the OID, treat the
small square (*place holder*) at its destination end, as if it were a node and
apply the *Reconnect* operation to make the change. For aesthetic reasons
it might be desirable to reposition the source end and this can also be
with the *Move* operation, described Section 4.1.1.

Outgoing connections also have a *place holder*, this time at the source
end. These connections can be transferred to an object using the
Reconnect operation as above and their destination ends can be placed
anywhere in the background with the *Move* operation. In both cases the
place holders disappear when they are disconnected

A1.4.2.3 Adding New External Connections

Because design is an inherently iterative process it cannot usually be
accomplished in a single top down pass. Thus situations arise in which
connections are sometimes overlooked at the high-level and the need for
them is often only detected when constructing a lower level OID. A provision is therefore made to remedy the situation from the bottom up, in

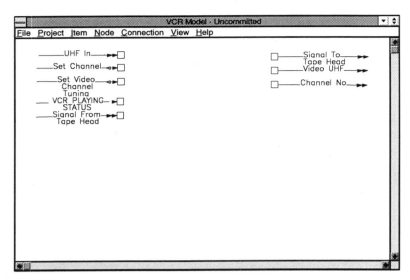

FIG. A1.6 External connections placed on new OID.

the form of an *Add Connection* operation on the *Connection Menu*. To add such a connection select *Add Connection* and position the source of the connection on an object in the outgoing case or the *background* in the incoming case, and then drag the end of the connection to the destination (object or background) and release the button. New external connections added in this way to an OID are automatically added to the parent diagram, but they will have a floating end that should be connected when the diagram is next edited. In the case of a CID a side effect of adding an external connection is that a matching entry is made in the associated CIS.

A1.4.2.4 Connections Involving Name Clashes

Several lines may be given the same label if they have the same definition. This might correspond to different instances of the same object or several objects having similar operations with identical names and parameters. Thus, two or more connections with the same name and type going to the same non-primitive object can ultimately be destined for different objects on lower level OIDs. This gives rise to the possibility of ambiguity in the diagrams involved. The editor allows such ambiguity to be resolved by extending the labels on the connections The extensions should prefix the labels with the names of the destination objects (followed by '::').

In the case where a new connection to a non-primitive object has the same name as one already present, the application will ask the user whether it is intended for the same final destination or not.

A1.4.2.5 Defining Nodes on CIDs

Three case occur in defining the nodes that appear on CIDs. The function and data store nodes are defined by C++ text. Inherited classes require no definition. Use of such a class must be consistent with its own definition that is located by the class name given in the node symbol. Finally control functions on CIDs are defined by state transition diagrams, application of the edit operation to such a node will produce initial diagram having one state, which can be completed in the normal way.

A1.5 TEXTUAL SPECIFICATIONS IN AN EXECUTABLE MODEL

In order to develop a Behavioural Model into an Executable Model, additional detail about the classes to which the primitive objects belong must be added. This consists of Class Interface Specifications (CISs), Class Implementation Diagrams (CIDs), Function Specifications (FSPECs) and Data Store Specifications (DSPECs). Also a mapping has to be given from the names used on OIDs into the names used in class definitions. The construction of CIDs has been outlined in Section 4 of this appendix, and this section focuses on the format required for the specification of textual information associated with primitive objects i.e. Mapping Specification, CISs, FSPECs and DSPECs.

A1.5.1 Matching Primitive Objects to Class Specifications

The behaviour of a primitive object in the OID structure is described by a MOOSE class definition. Class definitions are separate entities from the OID structure, and the class used by a primitive object is specified by a class name in the primitive object bubble. However, this name alone is not all that is required to allow a class to be utilised.

Additional information is contained in a textual OSPEC attached to the primitive object, which is created by applying the Descend option to an object with no children and then selecting the OSPEC option. An OSPEC consists of three distinct parts; a textual description of the primitive objects behaviour, a declaration of constant values, and the mapping of application specific names used in the OID hierarchy to C++ names used in the class. A summary of the structure of the OSPEC is presented below.

Place plain English text here ...
{
CONSTANTS:
Constant Name = String Replacement ; ... ;

NAME MAPPING:
Application specific name = Class name ; ... ;
Application constant name = Constructor parameter name ; ... ;
}

The textual English description is placed in the OSPEC outside the curly brackets, and is used as a comment describing the primitive objects behaviour. Information contained inside the brackets is used during the synthesis of an executable model and is described in further detail below.

The names used in a class description are selected so they are of local significance and help in the re-use of the class. In contrast to this the names used in the OID model are specific to the application and are selected to make the model more readable. Hence, a mapping is required to match the lines using application specific names on the OID model to the generic names used in the class.

A mapping of application specific names used in the OID model to names using C++ conventions that are present in the class description, is contained in the 'NAME MAPPING:' section of the OSPEC. Each application specific name is mapped to a class name with the '=' syntax. Multiple declarations can be separated by the ';' character.

Another important issue concerns the correct initialisation of the class. In many class definitions constructor function that accept parameters are defined, to allow different instances of a class to be generated from a single class description. In the OID model constants are declared which can then be passed to the constructor function of the class description during initialisation.

Constant declarations can be defined associated with a primitive object and these are supplied inside the OSPECs curly brackets and are prefixed by the 'CONSTANTS:' keyword. A constant's name can then be given a string substitution value using the '=' syntax and multiple constants can be declared by separating declarations by ';'. These constant declarations override any previous declarations of a constant higher up in the OID hierarchy (which are placed in declaration boxes).

Constant declarations need to be mapped onto constructor parameters and this mapping described in the NAME MAPPING section using a similar style to the mapping of flow names.

A1.5.2 Creating a Class Interface Specification (CIS)

A CIS defines the interface that the class provides to objects that communicate with it. This includes inputs such as the operations that the class provides, the events that it services and information flows that it accepts, and outputs such as the interactions that it expects other objects to provide and the information flows and events that it generates. A CIS is constructed as a text edit dialogue box, which should contain a set of statements that start with a keyword (or key phrase) and end with a ';'. A statement can specify several connections of the same type in which case they are separated by ','s. The keywords are given in uppercase thus: INTERACTION IN, INTERACTION OUT, EVENT IN, EVENT OUT, CONT INFO IN, CONT INFO OUT PARAM EVENT, DISCRETE INFO, CONSTRUCTOR, and TYPE.

Each connection specified in a CIS has a name, which should follow the normal C++ conventions and if it has parameters they must be specified by type names, which are themselves specified in TYPE statements. Events have no parameters thus they would be defined as follows:

EVENT IN: event_name1, event_name2, ... ;
EVENT OUT: event_name3, event_name4, ... ;

Interactions have a name, and parameters and result whose type is specified in parenthesis (the result being preceded by a ':') thus:

INTERACTION IN: Interaction_Name1(param_type1,
                 param_type2, ..., :result_typea), ..;
INTERACTION OUT: Interaction_Name1(param_type3,
                 param_type4, ...,:result_typeb), ...;

Information flows do not have parameters in the sense that interactions do but they might have fields whose type has to be specified as shown below:

CONT INFO IN: flow_name1 (info_type), ...;
CONT INFO OUT: flow_name2 (info_type), ...;
DISCRETE INFO: flow_name3 (info_type), ...;

Parameterised events are similar to interactions except they do not have results otherwise they are specified in a similar way:

PARAM EVENT: EVENT_NAME (Type_Name1, Type_Name2,
                  ...), ...;

Constructor functions are specified in a CIS, in case the objects are created dynamically and the user needs to now the parameters. The format is similar to an interaction with no result:

CONSTRUCTOR: name (param_type1 = default1, param_typ2 =
                  default2, ... , ...);

Any type names used defining the connections must be defined by a type

statement. This statement accepts any C++ type definition enclosed in
'{}'. Also a type name without a definition may be introduced (as an
opaque type) if the type information is not to be used directly within the
class definition. During the synthesis process an opaque type is replaced
by the definition given for the corresponding type in the CIS of an associated object,

TYPE: param_type1{C++ type definition},
                       field_type10 {C++type definition},
                       param_type2, ... ;

A1.5.3 Defining FSPECs

Functions are defined in FSPECs which are written in standard C++,
although they utilise certain built-in functions which are necessary to
facilitate model execution. These are summarised below:

- CAUSE
  Cause an event, or a parameterised event.
  ```
  CAUSE (EVENT_NAME);
     CAUSE (EVENT_NAME, (par1, par2, ... ));
  ```

- CALL
  Call to a function in another object i.e. an interaction
  ```
  result = CALL (Interaction_Name, (par1, par2,
     ... , ... ));
  ```

- GET/PUT
  Output/Input time continuous flows
  ```
  GET (Time_Continous_Flow_Name, var);
     PUT (Time_Continous_Flow_Name, var);
  ```

- SEND
  Write to a time discrete information flow
  ```
  SEND (Time_Discrete_Flow_name, var);
  ```

- CREATE/DCALL
  Create is used for the dynamic creation of objects. DCALL provides
  the mechanism to call the functions of a dynamically created object.
  Note that Ptr_Object is the pointer which references a dynamically
  created object.
  ```
  ptr = CREATE (Class_Name (par1, ... , ...));
     result = DCALL (Ptr_Object Method, ( par1,
     ..., ... ));
  ```

- LOG
  Adds user-defined messages to the log which is automatically created
  when the model is executed.
  LOG ("Text String");

A1.5.4 Defining DSPECs

DSPECs define the structure of the data stores found on a CID. They in
the appropriate C++ style and a copy of these definitions will appear in
the private part of the class definition.

APPENDIX 2

Ward–Mellor Model of the Mine Pump Control System

A2.1 THE TRANSFORMATION SCHEMA

Context Diagram

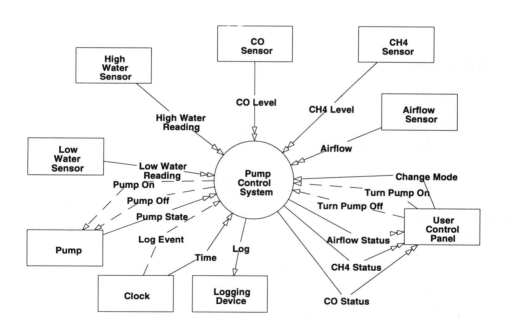

Level 0 Diagram

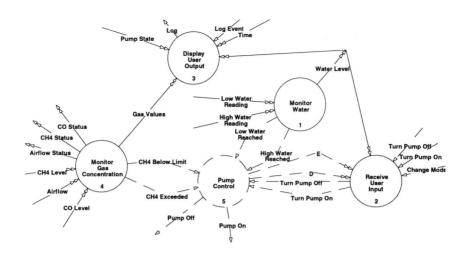

DFD Monitor Water

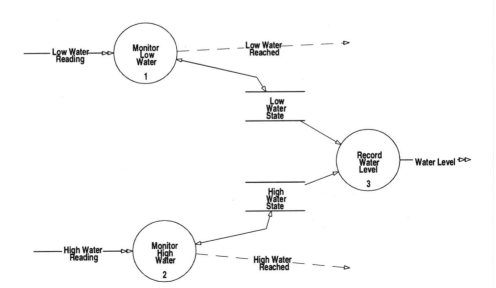

DFD Receive User Input

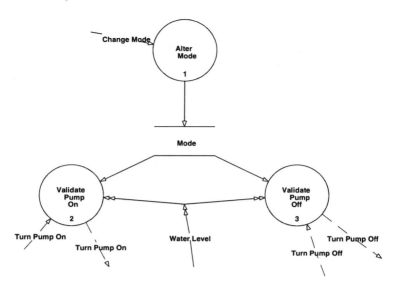

DFD Monitor Gas Concentration

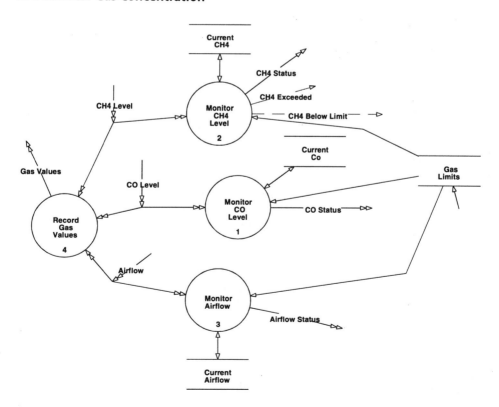

STD Pump Control

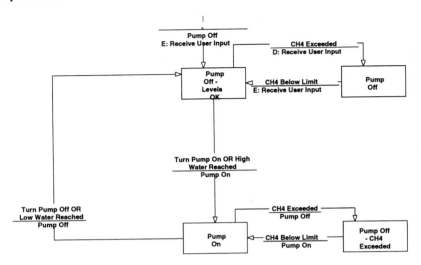

A2.2 DATA DICTIONARY

Air	airflow in m/sec
Airflow	airflow measurement from hardware
Airflow Limit	user set airflow safety level
Airflow Status	["SAFE"	"DANGER"] indicates if airflow is dangerous
CH4	CH4 level in ppm
CH4 Below Limit	CH4 level below safety threshold
CH4 Exceeded	CH4 level above safety threshold
CH4 Level	CH4 concentration measured by hardware
CH4 Limit	user set CH4 concentration safety level
CH4 Status	["SAFE"	"DANGER"] indicates if methane level is dangerous
Change Mode	["USER"	"SUPERVISOR"] Current operator status
CO	CO level in ppm
CO Level	CO concentration measured by hardware
CO Limit	user set CO safety limit
CO Status	["SAFE"	"DANGER"] indicates if CO level is dangerous
Current Airflow	["BELOW"	"ABOVE"] records airflow status
Current CH4	["BELOW"	"ABOVE"] records CH4 status
Current CO	["BELOW"	"ABOVE"] records CO status
Current Time	Logged time

Gas Limits	[CH4 Limit + CO Limit + Airflow Limit] safety values
Gas Values	[CH4 + CO + Airflow] current recorded gas values
High Water Reached	signals high water limit exceeded
High Water Reading	["HIGH"	"BELOW HIGH"] high water sensor reading
High Water State	["HIGH"	"BELOW HIGH"] records water level state
Level	["LOW"	"NORMAL"	"HIGH"] indicates water level state
Log	[CH4 + CO + Air + Current Time + Level + Pump] information logged
Log Event	Signal to instruct system to make new log entry
Log Info	[Gas Values	User Actions	Pump State] information for logging
Low Water Reached	signals high water limit exceeded
Low Water Reading	["LOW"	"ABOVE LOW"] low water sensor reading
Low Water State	["LOW"	"ABOVE LOW"] records low water status
Mode	["SUPERVISOR"	"USER"] records user status
Pump Off	signal turn off pump
Pump On	signal turn on pump
Time	indicates current Time
Turn Pump Off	user command requesting pump off
Turn Pump On	user command requesting turn pump on
Water Level	["LOW"	"NORMAL"	"HIGH"] indicates current water level

A2.3 PSPECS

PSPEC - Monitor Low Water

IF Low Water Reading = LOW AND Low Water State = ABOVE LOW
 Low Water Reached

Low Water State = LOW
END IF
IF Low Water Reading = NOT LOW AND Low Water State = LOW
  Low Water State = ABOVE LOW
END IF

PSPEC - Monitor High Water

IF High Water Reading = HIGH AND High Water State = BELOW HIGH
  High Water Reached
  High Water State = HIGH
END IF
IF High Water Reading = BELOW HIGH AND High Water State = HIGH
  Low Water State = BELOW HIGH
END IF

PSPEC - Record Water Level

IF Low Water State = LOW AND High Water State = ABOVE HIGH
  Water Level = LOW
END IF
ELSE IF Low Water State = ABOVE LOW AND High Water State =
ABOVE HIGH
  Water Level = NORMAL
END IF
ELSE IF Low Water State = ABOVE LOW AND High Water State = HIGH
  Water Level = HIGH
END IF

PSPEC - Alter Mode

IF Change Mode = SUPERVISOR AND Mode = USER
  Mode = SUPERVISOR
END IF
IF Change Mode = USER and Mode = SUPERVISOR
  Mode = USER
END IF

PSPEC - Validate Pump On

IF Mode = SUPERVISOR
  Turn Pump On
ELSE IF Water Level = NORMAL
  Turn Pump On

PSPEC - Validate Pump Off

```
IF Mode = SUPERVISOR
    Turn Pump Off
ELSE IF Water Level = BETWEEN LIMITS
    Turn Pump Off
```

PSPEC - Display User Output

```
Log.CH4 = Gas Value.CH4
Log.CO = Gas Value.CO
Log.Air = Gas Value.Air
Log.Current Time = Time
Log.Level = Water Level
Log.Pump = Pump State
```

PSPEC - Monitor CH4 Level

```
IF CH4 Level > CH4 Limit AND Current CH4 = BELOW
    Current CH4 = ABOVE
    CH4 Exceeded
END IF
IF CH4 Level <= CH4 Limit AND Current CH4 = ABOVE
    Current CH4 = BELOW
    CH4 Below Limit
END IF
```

PSPEC - Record Gas Values

```
Gas Values.Airflow = Airflow
Gas Values.CH4 = CH4 Level
Gas Values.CO = CO Level
```

PSPEC - Monitor CO Level

```
IF CO Level > CO Limit AND Current CO = BELOW
    Current CO = ABOVE
    Evacuate Status = EVACUATE
END IF
IF CO Level <= CO Limit AND Current CO = ABOVE
    Current CO = BELOW
    Evacuate Status = SAFE
END IF
```

PSPEC - Monitor Airflow

```
IF Airflow > Airflow Limit AND Current Airflow = BELOW
   Current Airflow = ABOVE
   Evacuate Status = EVACUATE
END IF
IF Airflow <= Airflow Limit AND Current Airflow = ABOVE
   Current Airflow = BELOW
   Evacuate Status = SAFE
END IF
```

APPENDIX 3

MOOSE Models for Mine Pump Control System

A3.1 THE BEHAVIOURAL MODEL

External View

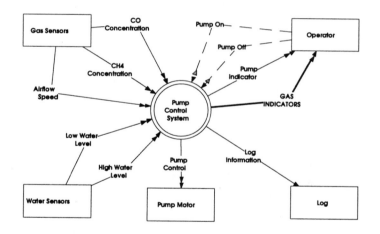

Functional View

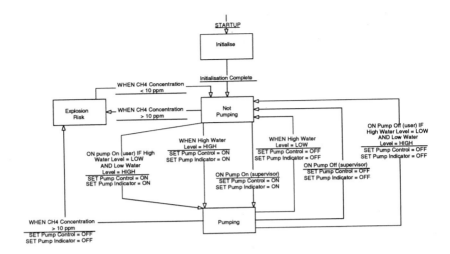

OID for Pump Control System

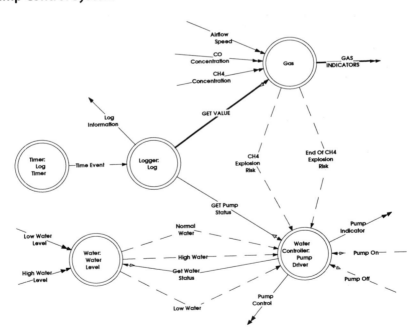

OID for Gas

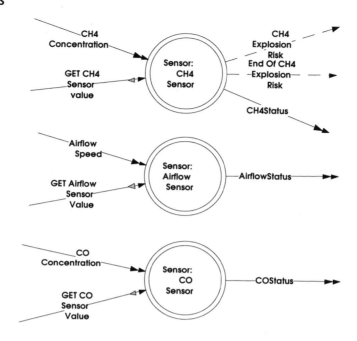

Model Dictionary

Airflow Speed	=	"Real Range 0-9" /*Analogue signal from airflow sensor*/
Airflow Status	=	"GasNormal | GasDanger"
Airflow Value	=	"Integer" /*Digitized speed of airflow in m/sec*/
CH4 Concentration	=	"Real Range 0-9 " /*Analogue signal from the CH4 sensor*/
CH4 Explosion Risk	=	/*CH4 level critical */
CH4 Status	=	"GasNormal | GasDanger"
CH4 Value	=	"Integer" /*Digitized concentration of CH4 in ppm*/
CO Concentration	=	"Real Range 0-9" /*Analogue signal from the CO sensor*/
CO Status	=	"GasNormal | GasDanger"
CO Value	=	"Integer" /*Digitized concentration of CO in ppm*/
date	=	day + month + year
day	=	"Integer Range 1 - 31"
End Of CH4 Explosion Risk	=	/*CH4 level is now safe*/
GAS INDICATORS	=	CH4 Status + CO Status + Airflow Status

GET Airflow Sensor Value	=	(:Airflow Value)
GET CH4 Sensor value	=	(:CH4 Value)
GET CO Sensor Value	=	(:CO Value)
GET Pump Status	=	(:Pump Status)
GET VALUE	=	GET CO Sensor Value + GET CH4 Sensor value + GET Airflow Sensor Value
GET Water Status	=	(:Water Status)
High Water	=	"Water level has gone above the high water sensor"
High Water Level	=	"Above | Below" /*Signal from the water sensor*/
hour	=	"Integer Range 0 - 23"
Log Information	=	CO Value + CH4 Value + Airflow Value + Time
Low Water	=	"Water level has fallen below the low water sensor"
Low Water Level	=	"Above | Below" /*Signal from water sensor*/
minute	=	"Integer Range 0 - 59"
month	=	"Integer Range 1 - 12"
Normal Water	=	/*Water level above low and below high water sensors*/
Operator	=	"Any | Supervisor"
Pump Control	=	"SwitchPumpOn | SwitchPumpOff"
Pump Indicator	=	"On | Off"
Pump Off	=	(Operator)
Pump On	=	(Operator)
Pump Status	=	"PumpOn | PumpOff"
second	=	"Integer Range 0 - 59"
Time	=	date + hour + minute + second
Time Event	=	/*Signals time to make log entry*/
Water Status	=	"High | Normal | Low" /*As shown by indicators*/
year	=	"Integer Range 0 - 99"

OSPECS

Log	This object creates a log entry containing the current time, gas levels and pump status when it receives a time event.
Water Level	This object generates events depending on the current water levels indicated by the

	water sensors. It can also be queried to return the current water status.
Log Timer	This object periodically generates time events.
Sensor	The class Sensor monitors gas concentration and flow, and signals dangerous and safe conditions. Three OSPECs would be defined for the CH4 Sensor, the CO Sensor and the Airflow Sensor.
Pump Driver	This object controls the water pump depending on the current water levels and whether the methane level is within operating limits. The object also allows the user to manually switch the pump on or off. The operation of the object is shown in the following table

	CH$_4$ Explosion Risk	End of CH$_4$ Explosion Risk	High Water	Low Water	Normal Water	Pump On (Supervisor)	Pump Off (Supervisor)	Pump On (User)	Pump Off (User)
Explosion Risk	—	Depends upon current water level	—	—	—	—	—	—	—
Low Off	Explosion Risk	—	High On		Normal Off	Low On	—	—	—
Low On	Explosion Risk	—	High On	—	Normal On	—	Low Off	—	—
Normal Off	Explosion Risk	—	High On	Low Off	—	Normal On	—	Normal On	—
Normal On	Explosion Risk	—	High On	Low Off	—	—	Normal Off	—	Normal Off
High Off	Explosion Risk	—	—	Low Off	Normal Off	High On	—	—	—
High On	Explosion Risk	—	—	Low Off	Normal On	—	High Off	—	—

A3.2 EXTENSIONS TO MAKE THE MODEL EXECUTABLE

Logger Class

Class mapping for object Log of class Logger

Log Information	= log_information
GET CH4 Sensor Value	= CH4_sensor_value

GET CO Sensor Value   = CO_sensor_value
GET Airflow Sensor Value   = airflow_sensor_value
GET Pump Status   = pump_status
Time Event   = time_event

CIS - Logger

EVENT IN: time_event;
INTERACTION OUT: CH4_sensor_value (:SensorValueType);
INTERACTION OUT: airflow_sensor_value (:SensorValueType);
INTERACTION OUT: CO_sensor_value (:SensorValueType);
INTERACTION OUT: pump_status (:PumpStatusType);
DISCRETE INFO: log_information (LogEntryType);

TYPE: SensorValueType, PumpStatusType, TimeType;
TYPE: LogEntryType {struct {TimeType Time; SensorValueType
                    CH4_status, Airflow_status, CO_status;
                    PumpStatusType Pump_status;}};

CID - Logger

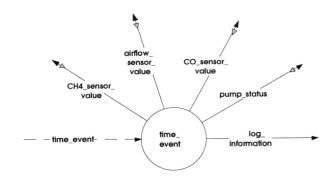

FSPECS - Logger

```
void time_event(EventType eid)
{
    LogEntryType log_entry;

    log_entry.Time = GET(simulated_time);
    log_entry.CH4_status=
        CALL(CH4_sensor_value,());
    log_entry.Airflow_status=
        CALL(airflow_sensor_value,());
    log_entry.CO_status =
```

```
      CALL(CO_sensor_value,());
log_entry.Pump_status =
      CALL(pump_status,());
  SEND(log_information,log_entry);
}
```

Sensor class

Class mapping for object CO Sensor of class Sensor

CH4 Status	=	gas_alarm_status
CH4 Explosion Risk	=	above_threshold
End Of CH4 Explosion Risk	=	below_threshold
CH4 Concentration	=	gas_sensor_signal
GET CH4 Sensor value	=	read_digitized_value

Class mapping for object Airflow Sensor of class Sensor

Airflow Status	=	gas_alarm_status
Airflow Speed	=	gas_sensor_signal
GET Airflow Sensor Value	=	read_digitized_value

Class mapping for object CO Sensor of class Sensor

CO Status	=	gas_alarm_status
CO Concentration	=	gas_sensor_signal
GET CO Sensor Value	=	read_digitized_value

CIS - Sensor

CONT INFO IN:	gas_sensor_signal (float);
CONT INFO OUT:	gas_alarm_status (AlarmStatusType);
EVENT OUT:	above_threshold, below_threshold;
INTERACTION IN:	read_digitized_value (:SensorValueType);

TYPE: AlarmStatusType {enum {GasNormal, GasDanger}};
TYPE: SensorValueType (int), ThresholdType (int);

CID - Sensor

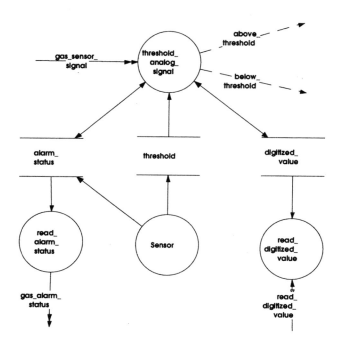

FSPECS - Sensor

```
void threshold_analog_signal()
{
    float signal;
    GET(gas_sensor,signal,signal);
    // Very roughly, emulate A-to-D with the
        following cast
    digitized_value = (SensorValueType)signal;
    if ((digitized_value > threshold)&&
        (alarm_status == GasNormal)) {
        alarm_status = GasDanger;
        CAUSE(above_threshold);
    }
    if ((digitized_value <
    threshold)&&(alarm_status == GasDanger)) {
        alarm_status = GasNormal;
        CAUSE(below_threshold);
    }
}
```

```
void read_alarm_status(){
   PUT(get_alarm_status, alarm_status);
}

SensorValueType read_digitized_value( ){
   return(digitized_value);
}

void Sensor (ThresholdType value){
   threshold = value;
   alarm_status = GasNormal;
}
```

DSPECS - Sensor

```
AlarmStatusType alarm_status;
ThresholdType threshold;
SensorValueType digitized_value;
```

Water Controller class

Class mapping for object Pump Driver of class Water Controller

Pump Indicator	=	pump_indicator
Pump On	=	pump_on
Pump Off	=	pump_off
Pump Control	=	pump_control
GET Pump Status	=	report_pump_status
CH4 Explosion Risk	=	explosion_risk
End Of CH4 Explosion Risk	=	end_of_risk
High Water	=	high_water
Low Water	=	low_water
Normal Water	=	normal_water
GET Water Status	=	current_water_status

CIS - Water Controller

EVENT IN:	CH4_explosion_risk, end_of_risk;
EVENT IN:	normal_water, low_water; high_water;
PARAM EVENT IN :	pump_off (OperatorType), pump_on (OperatorType);
CONT INFO OUT:	pump_indicator (BOOL), pump_control (PumpControlType);
INTERACTION IN :	report_pump_status (:PumpStatusType);

INTERACTION OUT: current_water_status (:WaterStatusType);
TYPE: WaterStatusType, PumpControlType {enum {SwitchPumpOff,
                         SwitchPumpOn}}
TYPE: PumpStatusType {enum {PumpOff, PumpOn}};
TYPE: OperatorType {enum{User, Supervisor}};
TYPE: PumpControlStates {enum {explode, hon, hoff, non, noff, lon,
                         loff, nil, undecided}};

CID - Water Controller

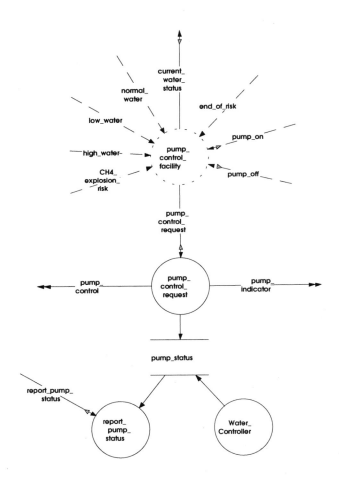

STDs - Water Controller

Pump Control Facility

The code for control functions is created automatically from STDs. In the diagram below the format of the transition reasons and responses have been simplified and the required code could not be synthesised.

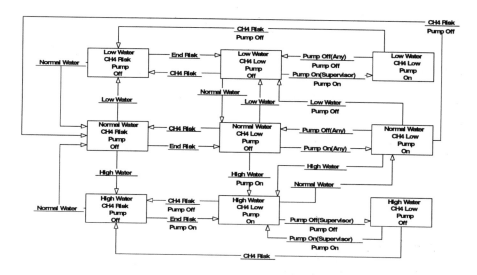

FSPECS - Water Controller

```
void pump_control_request(PumpStatusType
   switch_position){
   if (switch_position == PumpOn) {
      pump_status = PumpOn;
      PUT(pump_control,SwitchPumpOn);
      PUT(pump_indicator,On);
   }
   else{
      pump_status = PumpOff;
      PUT(pump_control,SwitchPumpOff);
      PUT(pump_indicator,Off);
   }
}


PumpStatusType report_pump_status(void){
   return (pump_status);
}
```

```
Water_Controller(void)
{
    pump_status = PumpOff;
}
```

DSPEC - Water Controller

PumpStatusType pump_status;

Water class

Class mapping for object Water Level of class Water

High Water Level	=	high_water_level
Low Water Level	=	low_water_level
Low Water	=	low_water
GET Water Status	=	water_status
High Water	=	high_water
Normal Water	=	normal_water

CIS - Water

CONT INFO IN:	high_water_level (WaterSensorSignalType);
CONT INFO IN:	low_water_level (WaterSensorSignalType);
EVENT OUT:	high_water, normal_water, low_water;
INTERACTION IN:	water_status (:WaterStatusType);

TYPE: WaterStatusType {enum {Low, Normal, High}};
TYPE: WaterSensorSignalType {enum {Below, Above}};

CID - Water

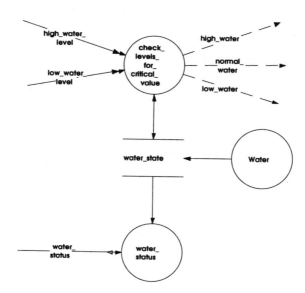

FSPECS - Water

```
void check_levels_for_critical_value()
{
   if ((high_water_level == Above) &&
   (water_state == Normal)) {
      water_state = High;
      CAUSE(high_water);
   }
   if ((high_water_level == Below) &&
   (water_state == High)) {
      water_state = Normal;
      CAUSE(normal_water);
   }
   if ((low_water_level == Below) &&
   (water_state == Normal)) {
      water_state = Low;
      CAUSE(low_water);
   }
   if ((low_water_level == Above) &&
   (water_state == Low)) {
      water_state = Normal;
      CAUSE(normal_water);
   }
}
```

```
WaterStateType water_status(void){
    return(water_state);
}

Water (void){
    water_state = Normal;
}
```

DSPEC - Water

WaterStateType water_state;

Timer Class

Class mapping for object Log Timer of class Timer

Time event   =   out;

CIS - Timer

EVENT OUT: out;

CID - Timer

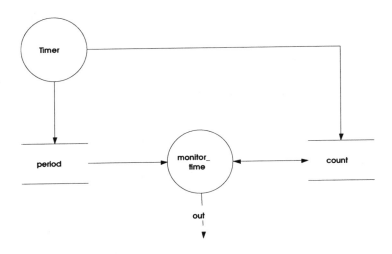

FSPECs - Timer

```
Timer (void){
    period = 300;
```

```
            count = 0;
        }

    void monitor_time(void){
        int current_time;
        current_time = GET(simulated_time);
        if ((current_time / period) != count) {
            if((current_time % period) == 0){
            count++; // ensures one event per tick
            CAUSE(out);
        }
    }
```

DSPECS - Timer

unsigned int period;
unsigned int count;

A3.3 THE COMMITTED MODEL

In the interests of brevity, the class definitions for the new objects are not shown here, though their development would follow the same pattern as previously described.

OID for Pump Control System - Committed

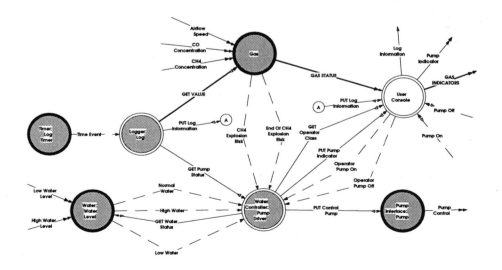

OID for Gas

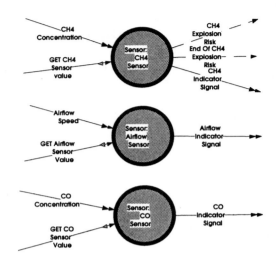

OSPECs

OSPECs for the new objects introduced by the commitment process are shown here; OSPECs for the other objects remain unchanged.

Pump Switch   Implements manual switching of pump
Pump   Interface object for controlling the pump
Formatter   Formats the log report
Security Lock   A two position keyed switch
User   A virtual object to map user desires onto interface.
                      For simulation purposes only, no
                      implementation required
Pump On Lamp   Lamp lit when pump on
CO Lamp   Lamp lit when CO concentration is dangerous
Airflow Lamp   Lamp lit when airflow is dangerous
CH4 Lamp   Lamp lit when CH4 concentration is dangerous
Printer   Object to print log

OID for Console

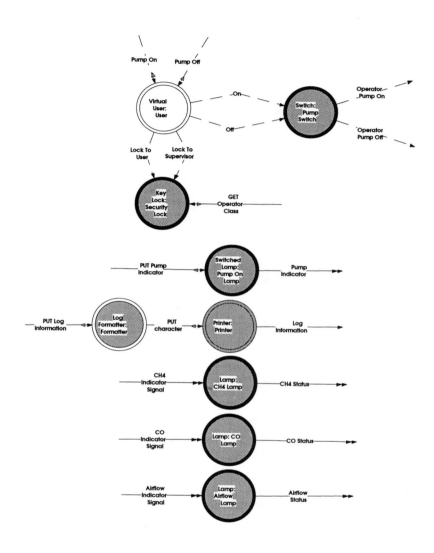

APPENDIX 4

VCR Control System

A4.1 THE BEHAVIOURAL MODEL

External View

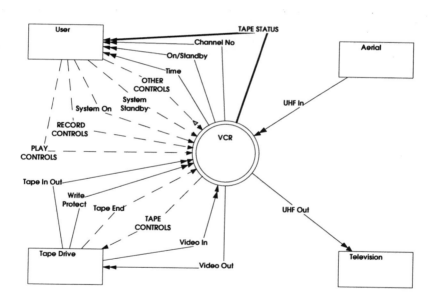

Functional View

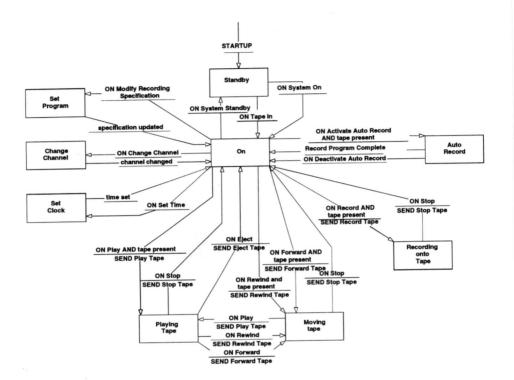

OID for VCR

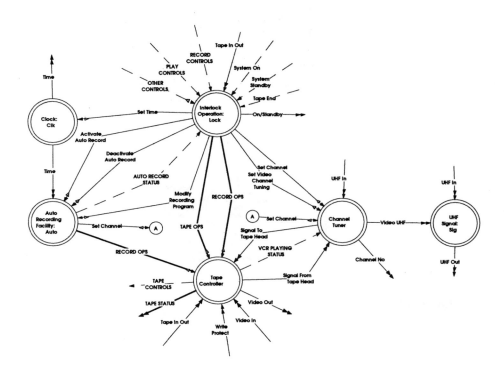

OID for Tape Controller

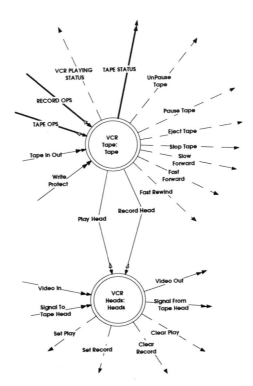

OID for Channel Tuner

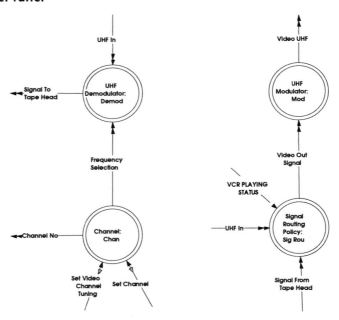

Model Dictionary

Activate Auto Record	=	()
AUTO RECORD	=	Auto Record Set + Auto Record Clear
Auto Record Clear	=	/* User generated timed programme abort */
Auto Record Finished	=	() /* All pending recording programs complete */
Auto Record Set	=	/* User generated timed programme start */
Auto Record Start	=	() /* Auto recording program started */
AUTO RECORD STATUS	=	Auto Record Finished+Auto Record Start+ Auto Record Stop
Auto Record Stop	=	() /*Auto recording program finished */
Change Channel	=	(channel)
channel	=	"Integer in range 1 - Max no channels" /* VCR tuning channel number */
Channel No	=	channel
Clear Play	=	/* Tape Drive to reset play heads */
Clear Record	=	/* Tape Drive to reset recording heads */
day	=	"Sun|Mon|Tue|Wed|Thu|Fri|Sat"
Deactivate Auto Record	=	()
Eject	=	/*User generated eject event */
Eject Op	=	()
Eject Tape	=	/* Tape Drive to eject tape */
Fast Forward	=	/* Tape Drive to transport tape forward at high speed */
Fast Rewind	=	/* Tape Drive to transport tape backward at high speed */
Forward	=	/* User generated forward event */
Forward Op	=	()
Forwarding	=	"True | False" /* Tape forwarding indicator */
frequency	=	"Integer in range Min tuning frequency - Max tuning frequency" /* UHF tuning frequency */

Frequency Selection	=	"Integer in range Max frequency – Min frequency" /* Demodulation frequency */
head position	=	"In | Out" /* virtual tape head position */
hour	=	"Integer in range 0 - 23"
minute	=	"Integer in range 0 - 59"
Modify Recording Program	=	(start time,stop time,channel)
Modify Recording Specification	=	(start time,stop time,channel)
New Time	=	(time) /* Change time of clock */
Not VCR Playing	=	/* Tape drive is not generating video event */
On/Standby	=	"On | Standby" /* Power indicator */
OTHER CONTROLS	=	New Time+Modify Recording Specification + Change Channel + Set Channel Tuning
Pause	=	/* User generated pause event */
Pause Op	=	()
Pause Tape	=	/* Tape Drive to stop tape transport */
Paused	=	"True | False" /* Tape pause indicator */
Play	=	/* User generated play event */
PLAY CONTROLS	=	Eject + Forward + Pause + Play + Rewind + Stop + Tape In
Play Head	=	(head position)
Play Op	=	()
Playing	=	"True | False" /* Tape playing indicator */
Record	=	/* User generated record event */
RECORD CONTROLS	=	Record + Stop + AUTO RECORD
Record Head	=	(head position)
Record Op	=	()
RECORD OPS	=	Record Op + Stop Op /* Operations required by the Auto Record facility */
Recording	=	"True | False" /* Tape recording indicator */
Rewind	=	/* User generated rewind event */
Rewind Op	=	()
Rewinding	=	"True | False" /* Tape rewinding indicator */
Set Channel	=	(channel)

Set Current Channel	=	(channel)
Set Play	=	/* Tape Drive to set heads to play from tape */
Set Record	=	/* Tape Drive to set heads to record to tape */
Set Time	=	(time)
Set Video Channel Tuning	=	(channel,frequency)
Set Video Tuning	=	(channel,frequency)
Signal From Tape Head	=	/* Video signal from playing operation */
Signal To Tape Head	=	/* Video signal for recording */
Slow Forward	=	/* Tape Drive to transport tape forward at play and record speeds */
start time	=	time
Stop	=	/* User generated stop event */
Stop Op	=	()
Stop Tape	=	/* Tape Drive to stop tape */
stop time	=	time
Stopped	=	"True | False" /* Tape stopped indicator */
System On	=	/* User generated on event */
System Standby	=	/* User generated off event */
TAPE CONTROLS	=	Eject Tape + Stop Tape + Slow Forward + Fast Forward + Fast Rewind + Set Play + Clear Play + Set Record + Clear Record + Pause Tape + UnPause Tape /* Commands to manipulate the tape */
Tape End	=	/*Tape end reached */
Tape In	=	/* User generated tape in event */
Tape In Out	=	"In | Out" /* Tape presence indication */
TAPE OPS	=	Forward Op + Rewind Op + Play Op + Pause Op + Record Op + Stop Op + Eject Op
Tape Present	=	"True | False" /* Tape present indicator */
TAPE STATUS	=	Paused + Playing + Recording + Forwarding + Rewinding + Stopped + Tape Present /* Indication to user about the status of the tape */
time	=	day + hour + minute

TV Signal	=	/* Picture from the demodulator containing the current TV picture */
UHF In	=	/* UHF signal from aerial */
UHF Out	=	/* UHF signal to the television */
UnPause Tape	=	/* Tape Drive to start tape transport */
VCR Playing	=	/* Tape drive is generating video event */
VCR PLAYING STATUS	=	VCR Playing+Not VCR Playing
Video In	=	/* Video signal from tape drive */
Video Out	=	/* Video signal to tape drive */
Video Out Signal	=	/* Video signal to be unmodulated */
Video UHF	=	/* UHF signal modulated locally */
Write Protect	=	"ReadWrite | ReadOnly" /* Indicates the write permission of the tape */

OSPECS

Demod	The Demod object takes the UHF in signal and demodulates it at the selected channel frequency to produce a raw video signal
Chan	The Chan object encapsulates the settings for the video channels, and these can be changed via the Set Video Channel Tuning event. The channel is changed by the Set Channel interaction, and the current channel is displayed to the user, via the Channel No dataflow. For the current video channel the object looks up the preset tuning and supplies it to the demodulator, via the Frequency Selection data flow.
Sig Rou	The Sig Rou object routes the demodulated TV signal from the aerial or the signal from the tape head to the Video Out Signal. The decision is made depending on the condition of the video. If the tape is playing (signified by the VCR Playing event) the Signal from Tape Head data is routed out. However, if the video is not playing (signified by the Not VCR Playing event) the TV Signal is routed out.
Sig	The Sig object merges the Video UHF signal (which contains the UHF signal generated by the video) with the UHF signal received from the Aerial.
Mod	This object takes the Video Out Signal picture and converts it into a UHF signal, called Video UHF.

Clk
This object continually provides the current time of day, in a 'day : hour : minute' format. The clock can be set to a value defined by the Set Time interaction.

Auto
This object is responsible for the management of the auto recording facility. Once activated the object determines the start and stop time of auto programs.

Tape
This object is responsible for the implementation of the set of basic tape operations.

Heads
The Heads object receives commands to move the play and record heads up and down via the play head and record head interactions respectively. The Set/Record Play and Record events are then issued that raise and lower the heads. The video in data signal is sent to the Signal from Tape Head data flow, which represents information off the play head. Similarly, the Signal from Tape Head data is routed to the Video out data flow, representing picture signals sent to the record head.

UserDisplay
This object is responsible for displaying basic tape status information

Timer
This object matches two times

Lock
This object implements the concurrency locking for the VCR system. At the highest level the system has four operating modes

- Standby - the idle, power saving mode
- On - the normal operating mode
- Waiting - a timed programme has been set and the system is awaiting the start of this programme - closely resembles the Standby mode
- Auto - a timed programme is being recorded - closely resembles the On mode

The transition between the modes is shown in the table below.

	On	Standby	Set Auto	Clear Auto	Auto Finished	Auto Start	Auto Stop	Stop
Standby	On	Standby	—	—	—	—	—	—
On	—	—	Waiting	—	—	—	—	Normal op
Waiting	—	—	—	On	—	Auto	—	—
Auto	—	Standby	—	—	Standby	—	Waiting	On

When in the On Mode, the tape operations that control the playing and recording of tapes can be undertaken. The modes, and the transitions between them, are shown in the table below.

	Tape In	Eject	Pause	Rewind	Record	Tape End	Play	Forward	Stop
Tape Out	Stopped	—	—	—	—	—	—	—	—
Stopped	—	Tape Out	—	Rewinding	Recording	—	Playing	Forwarding	—
Rewinding	—	—	—	—	—	Stopped	—	Forwarding	Stopped
Recording	—	—	—	—	—	Rewinding	—	—	Stopped
Playing	—	—	Paused	—	—	Rewinding	—	Previewing	Stopped
Forwarding	—	—	—	Rewinding	—	Rewinding	—	—	Stopped
Paused	—	—	—	—	—	—	Playing	—	Stopped
Previewing	—	—	—	—	—	—	Playing	—	Stopped

There are four operations concerned with configuring the VCR that can take place concurrently with the tape operations. These are:

- Setting the time
- Setting an Auto Recording Programme
- Changing a channel
- Setting the VCR's tuning

These operations are mutually exclusive.

A4.2 EXTENSIONS TO MAKE THE MODEL EXECUTABLE

In the interests of brevity, class mapping entries which transform English names into valid C names are not listed; for instance the entry Write Protect = write_protect.

Interlock Operation class

Class mapping for object Lock of class Interlock Operation

Auto Record Clear	=	auto_record_abort;
Modify Recording Program	=	programme_set;
Modify Recording Specification	=	set_programme
New Time	=	set_time;
On/Standby	=	on_off;
Set Channel	=	channel_change;
Set Time	=	time_set;
Set Channel Tuning	=	set_video_tuning;

Set Video Channel Tuning	=	video_tuning;
System On	=	on;
System Standby	=	standby;

CIS - Interlock Operation

EVENT IN :

> on, auto_record_set, auto_record_start, auto_record_abort, auto_record_stop, standby, auto_record_finished, tape_in, eject, pause, rewind, record, tape_end, play, forward, stop;

PARAM EVENT IN:

> change_channel(ChannelType), set_video_tuning(ChannelType, TuningFrequencyType), set_time(TimeType), set_programme(ProgrammeEntryType, ActiveProgrammeType);

CONT INFO IN :

> tape_in_out InOutType;

INTERACTION OUT :

> eject_op (), rewind_op (), record_op (:BOOL), play_op (),forward_op (), stop_op (), pause_op (BOOL), activate_record (BOOL), deactivate_record (BOOL) set_new_time(TimeType), set_tuning(ChannelType, TuningFrequencyType), modify_programme(ProgrammeEntryType, ActiveProgrammeType), set_channel(ChannelType);

TYPE : InOutType DEFINITION {enum{GoingOut,GoingIn}};
TYPE : SettingStatesType DEFINITION {enum {None,Prog, Channel,Time}};
TYPE : SystemStateType DEFINITION {enum {Standby,OnTapeIn,OnTapeOut, AutoWaiting,AutoRecording}};
TYPE : TapeControlStateType DEFINITION {enum {Out,Stopped, Rewinding,Recording,Playing,Forwarding, Paused,Previewing,Nil,Reset,Start}};
TYPE : OnStandType DEFINITION {enum {OST_Standby,On}};
TYPE : ChannelType, ProgrammeEntryType, ActiveProgrammeType, TuningFrequencyType,TimeType, DisplayStatesType;

CID - Interlock Operation

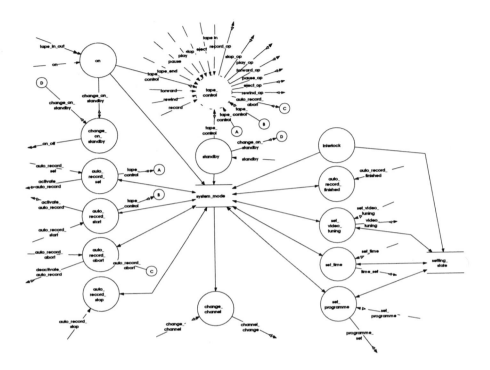

STD - Interlock Operation

The code for control functions is created automatically from STDs. In the diagram below the format of the transition reasons and responses have been simplified, and the required code could not be synthesised

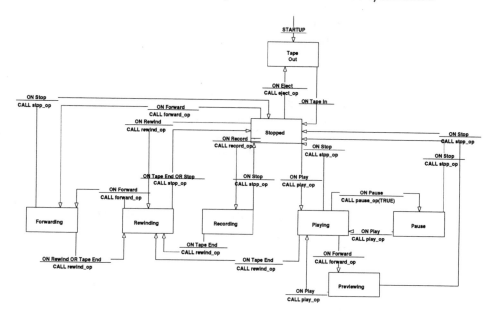

FSPECS - Interlock Operation

```
Interlock (void){
   system_mode = Standby;
   setting_state = None;
}

void standby(EVENT_TYPE eid){
   if ((system_mode == OnTapeIn) ||
   (system_mode == OnTapeOut) ||
      (system_mode == AutoRecording)){
      system_mode = Standby;
      change_on_standby(OST_Standby);
      CALL(tape_control,(Reset));
   }
}
```

```
void change_on_standby(OnStandType new_status){
    PUT(ON_STANDBY,new_status);
}

void auto_record_start(EVENT_TYPE eid){
    if(system_mode == AutoWaiting){
        system_mode = AutoRecording;
        CALL(activate_record,(TRUE));
    }
}

void auto_record_abort(EVENT_TYPE eid){
    if(system_mode == AutoRecording){
        system_mode = OnTapeIn;
        CALL(deactivate_record,(FALSE));
    }
}

void auto_record_finished(EVENT_TYPE eid){
    if(system_mode == AutoRecording) {
        system_mode = Standby;
        CALL(deactivate_record,(FALSE));
    }
}

void set_time(TimeType time){
    if((setting_state == None) && ((system_mode
    == OnTapeIn) || (system_mode ==
    OnTapeOut))){
        setting_state = Time;
        CALL(set_new_time,(time));
        setting_state = None;
    }
}

void set_video_tuning(unsigned int channel,float
    frequency) {
    if ((setting_state == None) && ((system_mode
    == OnTapeIn) || (system_mode ==
    OnTapeOut))){
        setting_state = Channel;
        CALL(set_tuning,(channel,frequency));
        setting_state = None;
    }
}
```

```
void auto_record_set(EVENT_TYPE eid){
   if(system_mode == OnTapeIn){
      system_mode = AutoWaiting;
      CALL(activate_record,(TRUE));
      tape_control(Reset);
   }
}

void auto_record_stop(EVENT_TYPE eid){
   // One programme has finished, 1 more to go
   if(system_mode == AutoRecording)
      system_mode = AutoWaiting;
}

void on(EVENT_TYPE eid){
   InOutType in_or_out;

   CALL(change_on_standby,(On));
   in_or_out = get(TAPE_IN_OUT);
   if(in_or_out == GoingIn){
      system_mode = OnTapeIn;
      CALL(tape_control,(TAPE_IN));
   }
   else
      system_mode = OnTapeOut;
}

void set_programme(ProgrammeEntryType
   programme,ActiveProgrammeType no){
   if((setting_state == None) && ((system_mode
   == OnTapeIn) || (system_mode ==
   OnTapeOut))){
      setting_state = Prog;
      CALL(modify_programme,(programme,no));
      setting_state = None;
   }
}

void change_channel(unsigned int channel_no){
   if((system_mode == OnTapeIn) || (system_mode
   == OnTapeOut))
      CALL(set_channel,(channel_no));
}
```

DSPECS - Interlock Operation

> SettingStatesType setting_state;
> SystemStateType system_mode;

User Display class

Class mapping for object UserDisplay of class User Display

PUT rewinding	=	rewinding;
PUT tape_present	=	tape_present;
PUT paused	=	paused;
PUT playing	=	playing;
PUT forwarding	=	forwarding;
PUT recording	=	recording;
PUT stopped	=	stopped;

CIS - User Display

> INTERACTION IN :   display void(DisplayStatesType);
> CONT INFO OUT :   paused BOOL, stopped BOOL, playing
>   BOOL, rewinding BOOL, forwarding
>   BOOL, recording BOOL, tape_present
>   BOOL;

> TYPE : DisplayStatesType DEFINITION {enum {Play,Rewind,Forward,
>   Pause,NotPause,Record,Stop,TapeIn,
>   TapeOut}};

CID - User Display

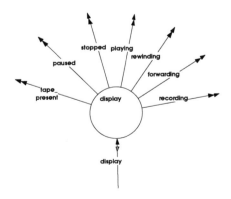

```
void display(DisplayStatesType toDisplay)
{
   switch (toDisplay) {
      case Play :
         PUT(Paused,FALSE);
         PUT(Stopped,FALSE);
         PUT(Playing,TRUE);
         break;
      case Rewind :
         PUT(Stopped,FALSE);
         PUT(Rewinding,TRUE);
         break;
      case Forward :
         PUT(Stopped,FALSE);
         PUT(Forwarding,TRUE);
         break;
      case Pause :
         PUT(Paused,TRUE);
         break;
      case NotPause :
         PUT(Paused,FALSE);
         break;
      case Record :
         PUT(Stopped,FALSE);
         PUT(Recording,TRUE);
         break;
      case Stop :
         PUT(Playing,FALSE);
         PUT(Rewinding,FALSE);
         PUT(Forwarding,FALSE);
         PUT(Paused,FALSE);
         PUT(Recording,FALSE);
         PUT(Stopped,TRUE);
         break;
      case TapeIn :
         PUT(Tape_present,TRUE);
         break;
      case TapeOut :
         PUT(Tape_present,FALSE);
         break;
   }
}
```

VCR Tape class

Class mapping for object Tape of class VCR Tape

Eject Op	=	eject;
Forward Op	=	forward;
Not VCR Playing	=	vcr_not_playing;
Pause Op	=	pause;
Play Op	=	play;
PUT Forwarding	=	forwarding;
PUT Paused	=	paused;
PUT Playing	=	playing;
PUT Recording	=	recording;
PUT Rewinding	=	rewinding;
PUT Stopped	=	stopped;
PUT Tape Present	=	tape_present;
Record Op	=	record;
Rewind Op	=	rewind;
Stop Op	=	stop;
Tape In Out	=	tape_in;

CIS - VCR Tape

CONT INFO IN :   tape_in InOutType, write_protect BOOL;

INTERACTION IN :   stop (), record (:BOOL), pause (BOOL),
play (), rewind (), forward (), eject (),
display (DisplayStatesType),
record_head(BOOL),
play_head(BOOL);

EVENT OUT :   not_vcr_playing, vcr_playing,
stop_tape, slow_forward, pause_tape,
unpause_tape, fast_rewind, eject_tape;

TYPE : InOutType;
TYPE : DisplayStatesType;

CID - VCR Tape

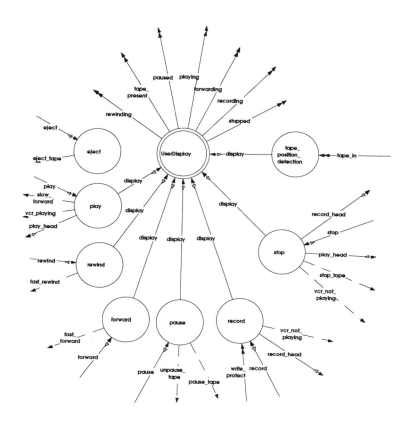

FSPECS - VCR Tape

```
void tape_position_detection(void){
  BOOL tape;

  tape = GET(tape_in);
  if (tape) CALL(display,(TapeIn));
  else CALL(display,(TapeOut));
}

void stop(void){
  CALL(record_head,(FALSE));
  CALL(play_head,(FALSE));
  CAUSE(vcr_not_playing);
  CAUSE(stop_tape);
  CALL(display,(Stop));
}
```

```
BOOL record(void){
  BOOL protection;

  protection = GET(write_protect);
  if(!protection){
      CAUSE(vcr_not_playing);
      CAUSE(slow_forward);
      CALL(record_head,(TRUE));
      CALL(display,(Record));
      return(TRUE);
  }
  else return(FALSE);
}
void pause(BOOL on){
  if(on){
      CAUSE(pause_tape);
      CALL(display,(Pause));
  }
  else{
      CAUSE(unpause_tape);
      CALL(display,(NotPause));
  }
}
void play(void){
  CAUSE(vcr_playing);
  CAUSE(slow_forward);
  CALL(play_head,(TRUE));
  CALL(display,(Play));
}
void rewind(void){
  CAUSE(fast_rewind);
  CALL(display,(Rewind));
}
void forward(void){
  CAUSE(fast_forward);
  CALL(display,(Forward));
}
void eject(void){
  CAUSE(eject_tape);
}
```

UHF Demodulator class

Class mapping for object Demod of class UHF Demodulator

Frequency Selection	=	demodulation_frequency;
Signal To Tape Head	=	demodulated_signal;
UHF in	=	modulated_signal;

CIS - UHF Demodulator

CONT INFO IN : modulated_signal UHFType,
demodulation_frequency
TuningFrequencyType;

CONT INFO OUT : demodulated_signal TVSignalType;

TYPE : UHFType DEFINITION float;
TYPE : TVSignalType DEFINITION {float};
TYPE : TuningFrequencyType;

CID - UHF Demodulator

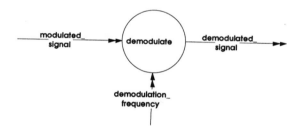

FSPECS - UHF Demodulator

```
void demodulate (void)
{
   UHFType in;
   TuningFrequencyType freq;
   TVSignalType out;

   in = GET(modulated_signal);;
   freq = GET(demodulation_frequency);
   out = analog_demodulate(in,freq);
   //Emulation of hardware
   PUT(demodulated_signal,out);
}
```

Signal Routing Policy class

Class mapping for object Sig Rou of class Signal Routing Policy

VCR Playing	=	set_path_a;
Not VCR Playing	=	set_path_b;
UHF In	=	signal_b;
Signal From Tape Head	=	signal_a;
Video Out Signal	=	output;

CIS - Signal Routing Policy

EVENT IN :   set_path_a, set_path_b;
CONT INFO IN :   signal_a UHFType, signal_b UHFType;
CONT INFO OUT :   output UHFType;

TYPE : PathType DEFINITION {enum{A,B}};
TYPE : UHFType;

CID - Signal Routing Policy

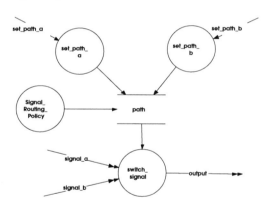

FSPECS - Signal Routing Policy

```
Signal_Routing_Policy(void) {
   path = A;
}
void set_path_a(EVENT_TYPE eid) {
   path = A;
}
void set_path_b(EVENT_TYPE eid) {
   path = B;
}
```

```
    void switch_signal(void) {
      UHFType a,b;
      if (path == A) {
        a = GET(signal_a);
        PUT(output,a);
      }
      if (path == B) {
        b = GET(signal_b) ;
        PUT(output,b);
      }
    }
```

DSPEC - Signal Routing Policy

    PathType Path;

Channel class

Class mapping for object Chan of class Channel

    CONSTANTS:
    FORMAT = 1;
    CLASS MAPPING:
    Frequency Selection   =   frequency;
    Set Video Channel Tuning   =   set_tuning;

CIS - Channel

    INTERACTION IN :   set_channel(ChannelType);
    INTERACTION IN :   set_tuning(ChannelType,
                                   TuningFrequencyType);
    CONT INFO OUT :   channel_no ChannelType;
    CONT INFO OUT :   frequency FrequencyType;
    TYPE : ChannelType DEFINITION {unsigned int};
    TYPE : TuningFrequencyType DEFINITION {float};
    TYPE : FrequencyType DEFINITION {unsigned long};

CID - Channel

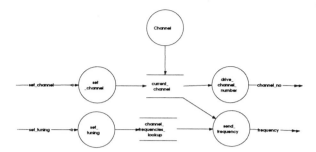

FSPECS - Channel

```
Channel(void)
{
   current_channel = 1;
}

void set_channel(ChannelType channel)
{
   if ((channel>0)&&(channel<=20))
      current_channel = channel;
   else current_channel = 20;
}

void set_tuning(ChannelType
   channel,TuningFrequencyType frequency)
{
   if ((channel>0) && (channel<=20))
      channel_frequency_lookup[channel-1] =
   frequency;
}

void send_frequency(void)
{
   FrequencyType display_freq;
   display_freq = (FrequencyType)
   (channel_frequency_lookup[current_channel-
   1]);
   PUT(frequency,display_freq);
}
```

```
void drive_channel_number(void)
{
    PUT(channel_no,current_channel);
}
```

DSPECS - Channel:

ChannelType current_channel;
TuningFrequencyType channel_frequencies_lookup[20];

UHF Modulator class

Class mapping for object Mod of class UHF Modulator

CONSTANTS:
MODFREQ = 148;
CLASS MAPPING:
Video UHF   =   signal_in;
Video Out Signal   =   signal_out;

CIS - UHF Modulator

CONT INFO IN :   signal_in TVSignal_Type;
CONT INFO OUT :   signal_out UHFType;

TYPE : TVSignalType;
TYPE : UHFType;

CID - UHF Modulator

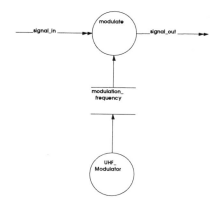

FSPECS - UHF Modulator

```
UHF_Modulator(float frequency){
   modulation_frequency = frequency;
}

void modulate (void) {
   UHFType in,out;

   in = GET(signal_in);
   out = analog_modulate(in,freq); // Emulation
   of H/W
   PUT(signal_out,out);
}
```

DSPEC - UHF Modulator

float modulation_frequency;

UHF Signal class

Class mapping for object Sig of class UHF Signal

Video UHF   =   video;

CIS - UHF Signal

CONT INFO IN :   video UHFType, uhf_in UHFType;
CONT INFO OUT :   uhf_out UHFType;
TYPE : UHFType;

CID - UHF Signal

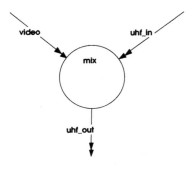

FSPECS - UHF Signal

```
void mix(void){
   UHFType video, standard, result;

   video = GET(video);
   standard = GET(uhf_in);
   result = video + standard;
   PUT(uhf_out, result);
   // acts as H/W emulator
}
```

Clock class

CIS - Clock

INTERACTION IN :	set_time(TimeType);
CONT INFO OUT :	time TimeType;

TYPE : TimeType DEFINTION {struct {DayType day;int hour,minute;}};
TYPE : DayType DEFINITION {enum {Sun,Mon,Tue,Wed,Thu,Fri,Sat}};

CID - Clock

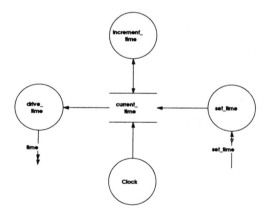

FSPEC - Clock

```
Clock(void) {
   current_time.day = Sun;
   current_time.hour = 0;
   current_time.minute = 0;
}

void set_time(TimeType new_time){
   current_time = new_time;
}

void drive_time(void){
   PUT(time, current_time);
}

void increment_time(void){
// Active function - increments the time
   current_time.minute ++;
   if (current_time.minute > 59){
      current_time.minute = 0;
      current_time.hour ++;
      if (current_time.hour > 23){
         current_time.hour = 0;
         if (current_time.day == Sat)
            current_time.day = Sun;
         else current_time.day++;
      }
   }
}
```

DSPEC - Clock

TimeType current_time;

Timer class

CIS - Timer

INTERACTION IN : match(TimeType,TimeType:BOOL);

TYPE : TimeType;

CID - Timer

FSPEC - Timer

```
BOOL match(TimeType start_time,TimeType
   current_time)
{
   BOOL isSame = TRUE;

   if (start_time.day   !=current_time.day)
   isSame=FALSE;
   if (start_time.hour !=current_time.hour)
   isSame=FALSE;
   if (start_time.minute!=current_time.minute)
   isSame=FALSE;

   return(isSame);
}
```

Auto Recording Facility class

Class mapping for object Auto of class Auto Recording Facility

```
CONSTANTS:
FORMAT = 1;
CLASS MAPPING:
Activate Auto Record        =    activate_record;
Auto Record Finished        =    auto_end_2;
Auto Record Start           =    auto_start;
Auto Record Stop            =    auto_end_1;
Deactivate Auto Record      =    deactivate_record;
Modify Recording Program    =    modify_program;
Record Op                   =    record;
Stop Op                     =    stop;
```

CIS - Auto Recording Facility

EVENT OUT :	auto_start, auto_end1, auto_end2;
CONT INFO IN :	time TimeType;
INTERACTION IN :	modify_program (ProgramEntryType), activate_record (BOOL), deactivate_record (BOOL);
INTERACTION OUT :	record (), stop (), set_channel (ChannelType);

```
TYPE : EntryType DEFINITION {struct {
                             DayType day1, day2;
                             int hour1, hour2;
                             int minute1, minute2;
                             int channel;
                             }};
TYPE: DayType DEFINITION {enum{Mon,Tue,Wed,Thu,Fri,Sat,Sun}};
TYPE: ActiveProgrammeType DEFINITION {enum{One,Two}};
TYPE: ProgrammeType DEFINITION {struct{
                             ProgrammeEntryType prog_entry[2];
                             ActiveProgrammeType which_active;
                             }};
TYPE: TimeType DEFINITION {struct{
                             DayType day;
                             int hour, minute;
                             }};
TYPE : ChannelType;
```

CID - Auto Recording Facility

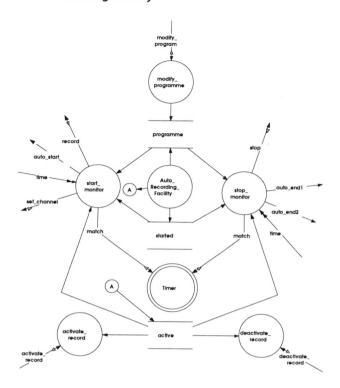

FSPECS - Auto Recording Facility

```
Auto_Recording_Facility (void){
    active = FALSE;
    started = FALSE;
    programme.programme_entry[0].day1 = Mon;
    programme.programme_entry[0].hour1 = 0;
    programme.programme_entry[0].minute1 = 0;
    programme.programme_entry[0].day2 = Mon;
    programme.programme_entry[0].hour2 = 0;
    programme.programme_entry[0].minute2 = 0;
    programme.programme_entry[0].channel = 0;
    programme.programme_entry[1].day1 = Mon;
    programme.programme_entry[1].hour1 = 0;
    programme.programme_entry[1].minute1 = 0;
    programme.programme_entry[1].channel = 0;
    programme.programme_entry[1].day2 = Mon;
    programme.programme_entry[1].hour2 = 0;
    programme.programme_entry[1].minute2 = 0;
```

```
      programme.programme_entry[1].channel = 0;
      programme.which_active = One;
}

void activate_record(BOOL on){
   if(on) {
      programme.which_active = One;
      active = TRUE;
   }
}

void start_monitor(void){
   TimeType current_time, start_time, end_time;
   int index;

   if(active && !started){
      current_time = get(time);
      if (programme.which_active == One)
         index = 0;
      else index = 1;
      start_time.day  =
         programme.programme_entry[index]
            .day1;
      start_time.hour  =
         programme.programme_entry[index].
            hour1;
      start_time.minute =
         programme.programme_entry[index].
            minute1;
      end_time.day  =
         programme.programme_entry[index].
            day2;
      end_time.hour  =
         programme.programme_entry[index].
            hour2;
      end_time.minute =
         programme.programme_entry[index].
            minute2;
```

```
        if (CALL(match,(start_time,
           current_time))){
           CALL(set_channel,(programme.
              programme_entry[index].channel));
           CAUSE(auto_start);
           CALL(record,());
        }
    }
}

void deactivate_record(BOOL off){
    if(!off) {
        active = FALSE;
        programme.which_active = One;
    }
}

void stop_monitor(void){
    TimeType current_time, start_time, end_time;
    int index;

    if(active && started){
        current_time = GET(time);
        if (programme.which_active == One)
           index = 0;
        else index = 1;
        end_time.day = programme.programme_entry
           [index].day2;
        end_time.hour = programme.programme_entry
           [index].hour2;
        end_time.minute = programme.
           programme_entry [index].minute2;
        if(CALL(match,(current_time, end_time))){
           CALL(stop,());
           programme.programme_entry[index].day1
              = Mon;
           programme.programme_entry[index].hour1
              = 0;
           programme.programme_entry[index].
              minute1 = 0;
           programme.programme_entry[index].day2
              = Mon;
           programme.programme_entry[index].hour2
              = 0;
```

```
            programme.programme_entry[index].
               minute2 = 0;
            programme.programme_entry[index].
               channel = 0;
            if(index == 0){
               // Is there a second programme?
               if (programme.programme_entry[1].
                  channel == 0) {
                  // Only one programme
                  CAUSE(auto_end2);
                  started = FALSE;
               }
               else {
                  CAUSE(auto_end1,NULL);
                  started = FALSE;
                  programme.which_active = Two;
               }
            }
            else{
               CAUSE(auto_end2D,NULL);
               programme.which_active = One;
            }
         }
      }
   }

   void modify_programme(ProgrammeEntryType prog,
      ActiveProgrammeType no){
      int index;

      if(no == One){
            index = 0;
      }
      else{
         index = 1;
      }
      programme.programme_entry[index].
         day1 = prog.day1;
      programme.programme_entry[index].
         hour1 = prog.hour1;
      programme.programme_entry[index].
         minute1 = prog.minute1;
      programme.programme_entry[index].
         day2  = prog.day2;
      programme.programme_entry[index].
```

```
        hour2   = prog.hour2;
programme.programme_entry[index].
    minute2 = prog.minute2;
programme.programme_entry[index].
    channel = prog.channel;
}
```

DSPECS - Auto Recording Facility

> BOOL active, started;
> ProgrammeType programme;

VCR Heads class

Class mapping for object Heads of class VCR Heads

Clear Play	=	not_playing;
Clear Record	=	not_recording;
Set Play	=	playing;
Set Record	=	recording;
Signal From Tape Head	=	from_heads;
Signal To Tape Head	=	to_heads;
Video In	=	signal_in;
Video Out	=	signal_out;

CIS - VCR Heads

INTERACTION IN :	play_head (BOOL), record_head void(BOOL);
CONT INFO IN :	signal_in UHFType,; to_heads UHFType, signal_out UHFType;
CONT INFO OUT :	from_heads UHFType;
EVENT OUT :	playing, not_playing, recording, not_recording;

> TYPE UHFType;
> TYPE RouteType DEFINITION {enum{PathPlaying,PathRecording, NoPath}};

CID - VCR Heads

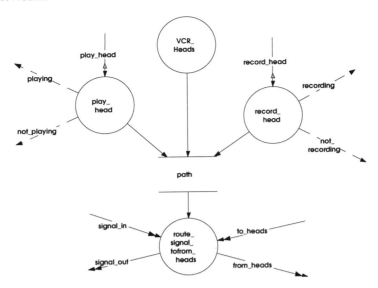

FSPECS - VCR Heads

```
VCR_Heads(void){
      path = NoPath;
}

void play_head(BOOL pl){
      if(pl){
          path = Playing;
          CAUSE(playing);
      }
      else{
          path = NoPath;
          CAUSE(not_playing);
      }
}

void route_signals_tofrom_heads(void){
    UHFType signal_to_heads,input;
        switch (path) {
            case PathPlaying :
                signal_to_heads = GET(to_heads);
                PUT(signal_out,signal_to_heads);
                PUT(from_heads,0.0);
                break;
```

```
          case PathRecording:
                  PUT(signal_out,0.0);
                  input = GET(signal_in);
                  PUT(from_heads,input);
                  break;
          case NoPath :
                  PUT(from_heads,0.0);
                  PUT(signal_out,0.0);
                  break;
      }
  }

  void record_head(BOOL rec){
     if(rec){
        path = Recording;
        CAUSE(recording);
     }
     else{
        path = NoPath;
        CAUSE(not_recording);
     }
  }
```

DSPEC - VCR Heads

RouteType path;

A4.3 THE COMMITTED MODEL

In the interests of brevity, the class definitions for the new objects are not
shown here, though their development would follow the same pattern as
previously described.

OID for VCR - Committed

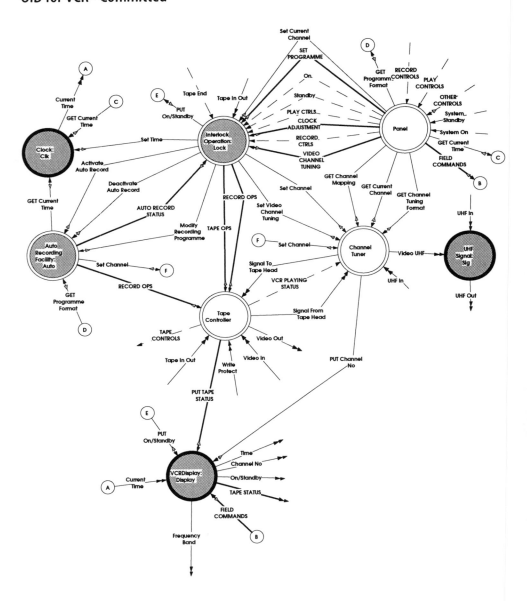

OID for Tape Controller - Committed

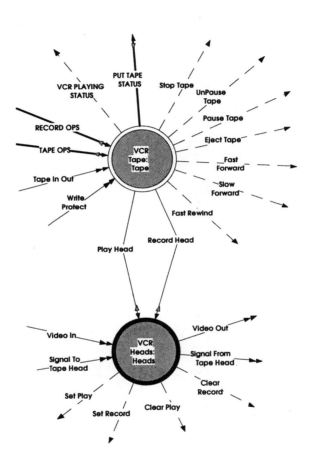

OID for Channel Tuner - Committed

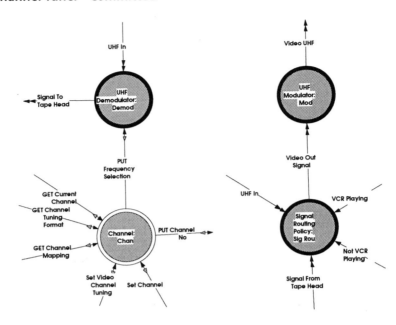

OID for Panel - Committed

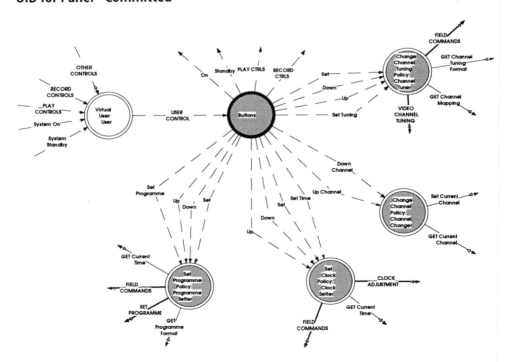

OID for Buttons - Committed

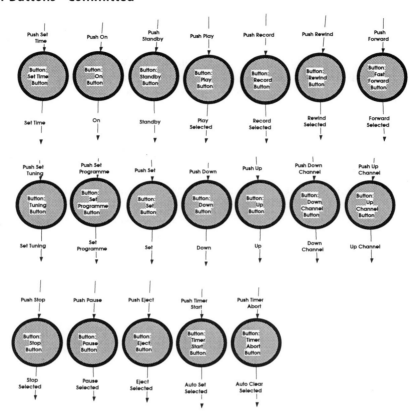

OSPECS

OSPECS for the new objects introduced by the committment process are shown here; OSPECS for the other objects remain unchanged.

Display	Gas plasma display for VCR
Time Display	Class used to display time information inherited into display class
Up Button	Button to increment selected item
Down Button	Button to decrement selected item
Set Button	Button to set selected item
Set Programme Button	Button to set programme
Set Time Button	Button to set time
Up Channel Button	Button to increment channel
Down Channel Button	Button to decrement channel
Tuning Button	Button to perform tuning

On Button	Button to switch VCR on
Standby Button	Button to switch VCR to standby
Play Button	Button to start VCR playing
Rewind Button	Button to start VCR rewinding
Fast Forward Button	Button to start VCR fast forwarding
Stop Button	Button to stop tape transport
Pause Button	Button to pause VCR
Eject Button	Button to eject tape
Timer Start Button	Button to start automatic programme recorder
Timer Abort Button	Button to abort automatic programme recorder
Channel Changer	Object to perform channel changing policy
Channel Tuner	Object to perform channel tuning policy
Clock Setter	Object to perform clock setting policy
Programme Setter	Object to perform programme setting policy

APPENDIX 5

Dynamic Object Creation

A5.1 THE BEHAVIOURAL MODEL

External

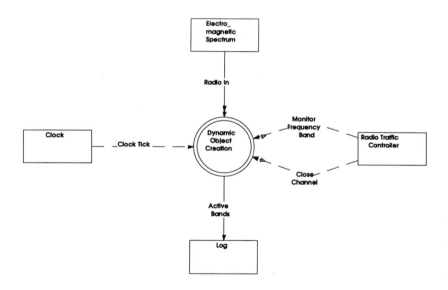

OID for Dynamic Model Creation

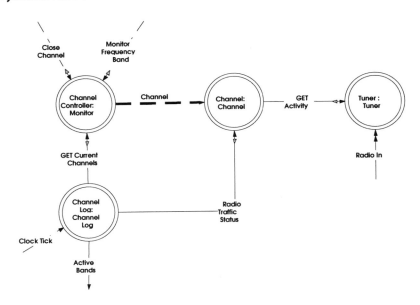

Model Dictionary

Active Bands	=	/*log entry for an active frequency band */
ActiveTunerFrequency	=	"Frequency"
Channel	=	Frequency "ChannelPointer" /*Creates instance of tuner class*/
Clock Tick	=	/*Event indicating one delta time step has elapsed*/
Close Channel	=	"Frequency_band" /*close channel indicated by frequency band*/
ChannelPointer	=	"Pointer to an active channel"
CurrentChannels	=	"ChannelPointer[10]"
GET Activity	=	(Frequency: TrafficPresent) /*Request to hardware tuner to determine if traffic is present at the frequency band*/
GET Current Channels	=	(:CurrentChannels)
Frequency	=	" Int "
Monitor Frequency Band	=	(Frequency_band) /*request to monitor a particular frequency*/
Radio In	=	/*Analogue waveforms*/
Radio Traffic Status	=	(:TrafficPresent) /* Method to

		return value indicating if traffic is present*/
TrafficPresent	=	"Present	Quiet" /* Frequency band activity status */

OSPECS

Channel	Child to monitor particular frequency
Monitor	Parent to make and delete child tuner objects.
Channel Log	Logs active channels
Tuner	Provides channel decoding at a specified frequency

A5.2 EXTENSIONS TO MAKE THE MODEL EXECUTABLE

Channel Class

Class mapping for object Channel of class Channel

GET Activity	=	activity
Channel	=	Channel
Radio Traffic Status	=	radio_traffic_status

CIS - Channel

INTERACTION IN:radio_traffic_status BOOL(void);
INTERACTION OUT: activity BOOL(int);

CID - Channel

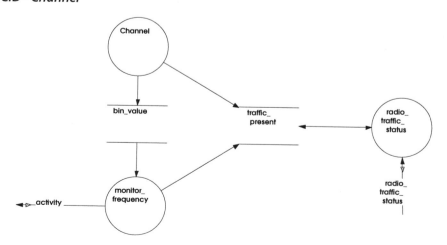

FSPECS - Channel

```
ChannelPointer Channel (int f){
    static ChannelPointer to_return;

    to_return = new TunerType(f);
    bin_value = f;
    traffic_present = FALSE;
    return(to_return);
}

void monitor_frequency(void){
    traffic_present = CALL(activity,(bin_value));
}

BOOL radio_traffic_status(void){
    BOOL toReturn;
    toReturn = traffic_present;
    traffic_present = False;
    return toReturn;
}
```

DSPEC - Channel

```
int bin_value;
BOOL traffic_present;
```

Channel Controller class

Class mapping for object Monitor of class Channel Controller

Monitor Frequency Band	=	monitor_frequency_band;
Close Channel	=	close_channel;
Channel	=	channel;
GET Current Channels	=	current_channels;

CIS - Channel Controller

```
DYNAMIC CONSTRUCTOR OUT:   Channel Channel(int);
INTERACTION IN:            current_channels void(TunerTableType *);
PARAM EVENT IN:            close_channel void(int);
PARAM EVENT IN:            monitor_frequency_band(int);

TYPE: ChannelPointerType;
```

TYPE: TunerTableType DEFINITION {struct{
  int iFreq;
  ChannelPointerType pTuner}};

CID - Channel Controller

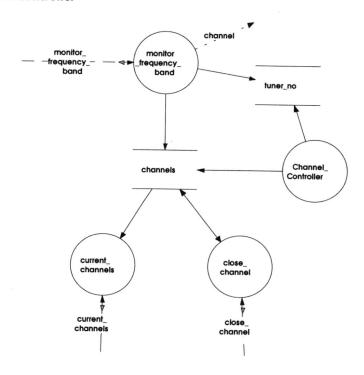

FSPECS - Channel Controller

```
Channel_Controller (void){
   int iTuner
   channels = (TunerTableType
   *)calloc(MAX_TUNERS,
   sizeof(TunerTableType));
   for (iTuner = 0; iTuner < MAX_TUNERS;
   iTuner++){
      (channels[iTuner]).iFreq = 0;
      (channels[iTuner]).pTuner = NULL
   }
   tuner_no = 0;
}
```

```
void monitor _Frequency_band(int iFreq){
   int iTuner = 0;

   BOOL bAdded = False;
   while((!bAdded) && (iTuner < MAX_TUNERS)){
      if((channels[iTuner]).pTuner == NULL){
         (channels[iTuner]).pTuner =
   CREATE(Tuner_class(iFreq));
         (channels[iTuner].iFreq = iFreq;
         bAdded = True;
      }
      tuner_no++;
   }
}

void close_channel(int iFreq)
{
   int iTuner=0;
   BOOL bRemoved = False;

   while ((!bRemoved) && (iTuner < MAX_TUNERS))
   {
      if ((channels[iTuner]).iFreq == iFreq) {
         channels[iTuner]).pTuner = NULL;
         channels[iTuner]).iFreq = 0;
         bRemoved=True;
      }
      iTuner++;
   }
   —tuner_no;
}

current_channels(TunerTableType
   *current_channels)
{
   int iTuner;

   for (iTuner = 0; iTuner < MAX_TUNERS;
   iTuner++) {
      (channels[iTuner]).pTuner =
   (current_channels[iTuner]).pTuner;
      (channels[iTuner]).iFreq =
   (current_channels[iTuner]).iFreq;
   }
}
```

DSPEC - Channel Controller

> TunerTableType * channels;
> int tuner_no

Channel Log class

Class mapping for object Channel Log of class Channel Controller

Clock Tick	=	clock_tick;
Active Bands	=	active_bands;
GET Current Channels	=	current_channels;
Radio Traffic Status	=	radio_traffic_status;

CIS - Channel Log

EVENT IN:	clock_tick;
INTERACTION OUT:	current_channels void(TunerTableType *);
DIS INFO OUT:	active_bands int;
DYNAMIC INTERACTION OUT:	radio_traffic_status pTunerClassType BOOL(void);

> TYPE: pTunerClassType;
> TYPE: TunerTableType;

CID - Channel Log

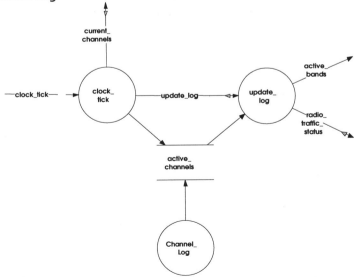

FSPECS - Channel Log

```
Channel_Log( void)
{
   active_channels = (TunerTableType
   *)calloc(MAX_TUNERS,
   sizeof(TunerTableType));
}

void clock_tick(void)
{
   CALL(current_channels(active_channels));
      CALL(update_log( ));
}

void update_log(void)
{
   int iTuner_counter;

   for (iTunercounter = 0; iTunercounter <
   MAX_TUNERS; iTunercounter++)
   if
   ((active_channels[iTuner_counter]).pTuner!=N
   ULL)
   if
   (DCALL((active_channels[iTuner_counter]).pTu
   ner.radio_traffic_status( )))
   SEND(active_bands,
   (active_channels[iTunercounter].iFreq);
}
```

DSPEC - Channel Log

```
int number_of_tuners
TunerTableType active_channels
```

Tuner Class

Class mapping for object Tuner of class Tuner

Radio In	=	radio_in;
GET Activity	=	activity;

CIS - Tuner

CONT INFO IN: radio_in float;
INTERACTION IN: activity BOOL(void);

CID - Tuner

FSPECS - Tuner

```
BOOL activity(void)
{
    // H/W emulation for simulation only
    if (radio_in>RADIO_THRESHOLD)
        return(TRUE);
    else return(FALSE);
}
```

REFERENCES

Abdel-Ghaly AA, Chan PY and Littlewood B (1986) 'Evaluation of Competing Software Reliability Predictions', *IEEE Trans. on Software Engineering*, Vol. SE-12, No. 9.

Agresti WM (1987a) 'What are the New Paradigms', in *New Paradigms for Software Development*, ed. Agresti, IEEE Computer Society Press, Los Angeles, USA, 6–10.

Agresti WM (1987b) 'A Case Study Extending Operational Specification to a Nonembedded System', in *New Paradigms for Software Development*, ed. Agresti, IEEE Computer Society Press, Los Angeles, USA, 179–186.

Agresti WM (1987c) 'Transformational Implementation', in *New Paradigms for Software Development*, ed. Agresti, IEEE Computer Society Press, Los Angeles, USA, 187–188.

Alavi M (1984) 'An Assessment of the Prototyping Approach to Information Systems Development', *Comm. ACM*, Vol. 27, No. 6, 556–563.

Ashworth C and Goodlang M (1990) *SSADM: A Practical Approach*, McGraw-Hill, New York, USA.

Audsley N, Burns A, Richardson M, Tindell K and Wellings AJ (1993) 'Applying new scheduling theory to static priority pre-emptive scheduling', *Software Engineering Journal*, Vol. 8 No. 5, 284–292.

Avizienis A (1985) 'The N-version approach to fault-tolerant software', *IEEE Trans. on Software Engineering*, Vol. SE-11, No. 12.

Balzer R (1981) 'Transformational Implementation: An Example', *IEEE Trans Soft Eng*, Vol. SE-7, No. 1, 3–14.

Barroca LM and McDermid JA (1992) 'Formal Methods : Use and Relevance for the Development of Safety-Critical Systems', *The Computer Journal*, Vol. 35, No. 6, 1992, 579–599.

Bate G (1986) 'MASCOT3: An Informal Introductory Tutorial', *Software Engineering Journal*, Vol. 1, No. 3, 95–102.

Bauer FL (1982) 'From Specifications to Machine Code: Program Construction through Formal Reasoning', *Proc. 6th International Conference on Software Engineering*, IEEE Press, Los Angeles, USA, 84–91.

Beason D (1995) 'Back to the Moon – At Last! The Clementine Solution', *Analog*, Vol. CXV, No. 10, 95–103.

Bland MJ and Evans DG (1994) 'Extending a high level modelling notation to model analogue subsystems', *2nd International Conference on Concurrent Engineering & Electronic Design Automation*, University of Bournemouth, The Society for Computer Simulation, 247–252.

Blumofe R and Hecht A (1988) 'Extending Real-Time Structured Analysis Specification', *ACM Sigsoft Soft. Eng. Notes*, Vol. 13, No. 3, 32–41.

Boehm BW (1976) 'Software Engineering', *IEEE Trans. On Computers*, Vol. C-25, No. 12, 1226–1241.

Boehm BW (1981) *Software Engineering Economics*, Prentice-Hall, Englewood Cliffs, NJ, USA.

Boehm BW (1988) 'A Spiral Model of Software Development and Enhancement', *IEEE Computer*, Vol. 21, No. 5, 61–72.

Boehm BW, Gray TE and Seewaldt T (1984) 'Prototyping Versus Specifying: A Multiproject Experiment', *IEEE Trans. Soft. Eng.*, Vol. SE-10, No. 3, 290–302.

Booch G (1991) *Object Oriented Design with Applications* Benjamin/Cummings, Redwood City, Ca, USA.

Booch G (1994) *Object Oriented Analysis and Design with Applications*, Benjamin/Cummings, Redwood City, Ca, USA.

Brooks FP (1987) 'No Silver Bullet : Essence and Accidents of Software Engineering', *IEEE Computer*, Vol. 20, No. 4, 10–19.

Burns A (1991) 'Scheduling Hard Real-time Systems: Review', *Software Engineering Journal*, Vol. 6, No. 3, 116–128.

Burns A and Lister AM (1991) 'A Framework for Building Dependable Systems', *The Computer Journal*, Vol. 34, No.2, 173–181.

Burns A and Wellings AJ (1989) *Real-Time Systems and their Programming Languages*, Addison-Wesley, Wokingham, UK.

Burns A and Wellings AJ (1994) 'HRT-HOOD : A Structured Design Method for Hard Real Time Systems', *Journal of Real Time Systems*, 6, 73–114.

Burns A, Tindell K and Wellings AJ (1995) 'Effective Analysis for Engineering Real-time Fixed Priority Schedulers', *IEEE Transactions on Software Engineering*, Vol. 21, No. 5, 475–480.

Burns A, Wellings AJ, Bailey CM and Fife E (1993) 'The Olympus Attitude and Orbital Control System : A Case Study in Hard Real-Time System Design and Implementation', Ada sans frontiers, *Proc. 12th Ada-Europe Conference*, Lecture Notes in Computer Science, Springer-Verlag.

Burns A, Wellings AJ, Hutcheon AD and Pierce RH (1994) 'Dependable Software Systems using Concurrency', Ada User Journal, Vol. 15, No. 2, 77–85.

Burton K, Miezejewski B, Roy J and Sheehan K (1991) 'Automatic Generation of Simulation Models from Structured Analysis', *Proc. 23rd Annual Summer Computer Simulation Conference*, Baltimore, MD.

Byte (1995) 'Notorious Bugs', *Byte*, Vol. 20, No. 9, Sept 1995, pp126.

Cameron JR (1986) 'An Overview of JSD', *IEEE Trans. Soft. Eng.*, Vol. SE-12, No.2, 222–240.

Chung L, Nixon B and Yu E (1995) 'Using Non-Functional Requirements to Systematically Support Change', *Proc. 2nd IEEE International Conference on Requirements Engineering*, York, UK, IEEE Press, 132–139.

Coad P and Yourdon E (1991) *Object-Oriented Analysis*, 2nd Edition, Prentice-Hall, Englewood Cliffs, NJ, USA.

Coleman D, Arnold P, Bodoff S, Dollin C, Gilchrist H, Hayes F and Jelemaes P (1991) *Object Oriented Development: The Fusion Method*, Prentice-Hall, Englewood Cliffs, NJ, USA.

D'Ambrosio JG and Hu X (1994) 'Configuration Level Hardware–Software Partitioning for Real Time Embedded Systems', *Proc Third International Workshop on Hardware/Software Codesign*, IEEE Press, Los Angeles, USA.

Davies J, Jackson D, Reed G, Reed J, Roscoe A and Schneider S (1992) 'Timed CSP: Theory and Practice', *Lecture Notes in Computer Science*, Vol. 600, Springer-Verlag, 640–675.

DeMarco T (1978) *Structured Analysis and System Specification*, Prentice-Hall, Englewood Cliffs, NJ, USA.

Dillinger TE (1988) *VLSI Engineering*, Prentice-Hall International Inc, 1988.

Easteal C and Davies G (1989) *Software Engineering Analysis and Design*, McGraw-Hill, New York, USA.

Ernst R, Henkel J and Benner T (1993) 'Hardware–Software Cosynthesis for Microcontrollers', *IEEE Design and Test of Computers*, Vol. 10, No. 4, 64–75.

European Commission (1991) *Background Material to the 1991 ESPRIT Work Programme*, Brussels, European Commission DGXIII.

Fidge CJ and Lister AM (1992) 'Disciplined Approach to Real-Time System Design', *Information and Software Technology*, Vol. 34, No. 9, 603–610.

Gane C and Sarson T (1979) *Structured Systems Analysis: Tools and Techniques*, Prentice-Hall, Englewood Cliffs, NJ, USA.

Gerber R, Hong S and Saksena M (1995) 'Guaranteeing Real-Time Requirements with Resource-Based Calibration of Periodic Processes' *IEEE Trans. on Software Engineering*, Vol. 21, No. 7, 579–592.

Ghezzi C, Mandriolo D, Morasca S and Pezze M (1989) 'A general way to put time in Petri Nets', in Proceedings of the Fifth International Workshop on Software Specification and Design, *ACM Sigsoft*, 60–67.

Gibbs WW (1994) 'Software's Chronic Crisis', *Scientific American*, Sept 1994, 72–81.

Gomaa H and Scott DBH (1981) 'Prototyping as a Tool in the Specification of User Requirements', *Proc. 5th International Conference on Software Engineering*, IEEE Press, Los Angeles, USA, 333–342.

Gupta RK and De Micheli G (1993) 'Hardware–Software Cosynthesis for Digital Systems', *IEEE Design and Test of Computers*, Vol. 10 No. 3, pp. 29–41.

Halfhill TR (1995) 'The Truth Behind the Pentium Bug', *Byte*, Vol. 20, No. 3, March 1995, 163–4.

Harel D (1987) 'Statecharts, A Visual Formalism for Complex Systems', *Science of Computer Programming*, Vol. 8, No. 3, 231–274.

Harel D (1992) 'Biting the Silver Bullet', *IEEE Software*, Vol. 25, No. 1, 8–20.

Harel D, Lachover H, Naamad A, Pnueli A, Polti M, Shermane R, Shtull-Trauring A and Trakhtenbrot M (1990) 'STATEMATE: A Working Environment for the Development of Complex Reactive Systems', *IEEE Trans. Software Eng.*, Vol. 16., No. 4, 403–414.

Hatley DJ and Pirbhai IA (1987) *Strategies for Real-Time System Specification*, Dorset House Publishing, New York, USA.

Hewlett-Packard (1989) *HP Teamwork User's Manual*, Hewlett-Packard, USA.

Hull MEC and O'Donoghue PG (1994) 'Timed Petri net approach to performance modelling with the MOON method', *Software Engineering Journal*, Vol. 9, No. 3, 95–106.

Jackson MA (1983) *Software Development*, Prentice Hall, Englewood Cliffs, NJ, USA

Jacobson I, Christerson M, Jonsson P and Övergaard G (1992) *Object-Oriented Software Engineering: A use case driven approach*, Addison-Wesley.

Jain PP, Dhinga S and Browne JC (1989) 'Bringing Top down Synthesis into the Real World', *High Performance Systems*, Vol. 10, No. 7, 86–94.

Jain R (1991) *The Art of Computer Systems Performance Analysis*, Wiley, 1991.

Kant K (1992) *Introduction to Computer System Performance Evaluation*, McGraw-Hill, New York, USA.

Kronolf K, Sheehan A and Hallmann M (1993) 'The Concept of Method Integration, in *Method Integration: Concepts and Case Studies*, ed. K. Kronolf, Wiley, New York, USA.

Kumar S, Aylor JH, Johnson BW and Wulf WA (1993) 'A Framework for Hardware/Software Codesign', *IEEE Computer*, Vol. 26, No. 12, pp. 39–45.

Lavi JZ, Agrawala AK, Buhr R, Jackson K, Jackson M and Lang B (1991) 'Formal establishment of computer-based systems engineering field urged', *IEEE Computer*, Vol. 24, no. 3, 105–107.

Lawrence PD and Mauch K (1988) *Real-Time Microcomputer System Design: An Introduction*, McGraw-Hill, 1988.

Leduc G (1991) 'An upward compatible timed extension to LOTOS', in K. Parker and G. Rose, eds, *Proceedings of FORTE'91, Fourth International Conference on Formal Description Techniques*, Elsevier Science, Publishers B.V. (North-Holland).

Leveson NG (1986) 'Software Safety: Why, What, and How', *ACM Computing Surveys*, Vol. 18, No. 2.

Liskov B, Snyder A, Atkinson R and Schaffert c (1977) 'Abstraction Mechanisms in CLU', *Communications of the ACM*, Vol. 20, No. 8, 564–576.

Littlewood B and Miller D (eds) (1991) *Software Reliability and Safety*, Elsevier Applied Science, London, UK.

Mealy GH (1955) 'A Method for Synthesising Sequential Circuits', *The Bell System Technical Journal*, Vol. 34, 1045–1079.

Mylopoulos J, Chung L and Nixon B (1992) 'Representing and Using Non-Functional Requirements: A Process-Oriented Approach', *IEEE Trans. on Software Engineering*, Vol. 18, No. 6, 483–497.

OOPSLA (1994) 'Methodology Standards: Help or Hinderance?', Panel 4, Ninth Annual Conference on Object-Oriented Programming Systems, Languages and Applications, *ACM Sigplan Notices*, Vol. 29, No. 10.

Parnas DL, Schouwen J and Kwan SP (1990) 'Evaluation of Safety-Critical Software', *Comm. ACM*, vol. 33, no. 6, 636–648.

Partsch H and Steinbrüggen R (1983) 'Program Transformation Systems', *ACM Computing Surveys*, Vol. 15, No. 3, 199–236.

Randell B (1975) 'System Structures for Software Fault Tolerance', *IEEE Trans. on Software Engineering*, Vol. SE-1, No.2, 220–232.

Robinson PJ (1992) *Hierarchical Object Oriented Design*, Prentice-Hall.

Royce WW (1970) 'Managing the Development of Large Software Systems: Concepts and Techniques', In *Proc. WESCON*, Section A1, 1–9, August 1970.

Rozenblit J and Buchenrieder K (eds.) (1995), *CODES: Computer Aided Software Hardware Design*, IEEE Press, Los Angeles, USA.

Rozier M, Abrossimov V, Armand F, Bowle I, Giln M, Guillemont M, Herruan F, Kaiser C, Langlois S, Leonard P and Newhauser W (1988) 'CHORUS Distributed Operating Systems', *Computer Systems Journal*, Vol. 1 No. 4.

Rumbaugh J, Blaha M, Pemerlani W, Eddy F and Lorensen W (1991) *Object-Oriented Modeling and Design*, Prentice-Hall, Englewood Cliffs, NJ, USA.

Schefstroem D and van den Broek G (eds.) (1993), *Tool Integration: Environments and Frameworks*, Wiley, New York, USA.

SELECT (1995) *Yourdon SELECT*, SELECT Software Tools Ltd, Cheltenham, Glos., England.

SES (1992) *SES/workbench User's Manual*, Scientific and Engineering Software.

Sha L, Rajkumar R and Lehoczky J (1990) 'Priority Inheritance Protocols: An Approach to Real-time Synchronisation', *IEEE Transactions on Computers*, Vol.39, No. 9, 1175–1185.

Shaw M, Wulf WA and London RL (1977) 'Abstraction and Verification in Alphard: Defining and Specifying Interaction and Generators', *Communications of the ACM*, Vol. 20, No. 8, 553–564.

Shlaer S and Mellor SJ (1992) *Object Lifecycles: Modelling the World in States*, Yourdon Press, Prentice-Hall, Englewood Cliffs, NJ, USA.

Simpson H (1986) 'The Mascot Method', *Software Engineering Journal*, Vol. 1, No. 3, 103–120.

Simpson JA and Weiner ESC (1989) *The Oxford English Dictionary*, Second Edition, Clarendon Press, Oxford, UK.

Sommerville I (1996) *Software Engineering* Addison-Wesley.

Srivastava MB and Brodersen RW (1992) 'Using VHDL for High-Level, Mixed-Mode System Simulation', *IEEE Design and Test of Computers*, Vol. 9 No. 3, 31–40.

Stallings W (1993) *Local and Metropolitan Area Networks*, MacMillan, Basingstoke, UK.

Stankovic JA (1988) 'Real-Time Computing Systems: The Next Generation', in *Hard Real-Time Systems*, Stankovic and Ramamritham (eds) IEEE Press, Los Angeles, USA.

Stroustrup B (1991) *The C++ Programming Language*, 2nd Edition, Addison Wesley.

Structured Software Systems (1994) *Cradle: Systems Engineering Guide*, Structured Software Systems Limited.

Swartout W and Balzer R (1982) 'On the Inevitable Intertwining of Specification and Implementation', *Comm. ACM*, Vol. 25., No. 7, 438–440.

Taylor T and Standish TA (1982) 'Initial Thoughts on Rapid Prototyping Techniques', *ACM SIGSOFT Software Engineering Notes*, Vol. 7, No. 5, 160–166.

Thomas DE, Adams JK and Schmit H (1993) 'A Model and Methodology for Hardware–Software Codesign', *IEEE Design and Test of Computers*, Vol. 10, No. 3, 6–15.

Thomé B (1993), 'Definition and Scope of Systems Engineering', in *Systems Engineering: Principles and Practice of Computer-Based Systems Engineering*, ed.

B. Thomé, Wiley, 1993.

Ward PT (1986) 'The Transformation schema: An Extension of the Data flow Diagram to Represent Control and Timing', *IEEE Trans. Soft. Eng.*, Vol. 12, No. 2, 198–210.

Ward PT (1989) 'How to Integrate Object Orientation with Structured Analysis and Design', *IEEE Software*, Vol. 6, No. 2, 74–81.

Ward PT and Mellor SJ (1985) *Structured Development for Real-Time Systems*, Vols 1, 2 and 3, Yourdon Press, Prentice-Hall, Englewood Cliffs, NJ, USA.

Weinburg V (1978) *Structured Analysis*, Yourdon Press, Prentice-Hall, Englewood Cliffs, NJ, USA.

Wirfs-Brock R and Wilkerson B (1989) 'Object-Oriented Design: A Responsibility-Driven Approach', In *OOPSLA '89 Proceedings*, *Sigplan* Notices, Vol. 24, No. 10, 71–76.

Woo NS, Dunlop AE and Woolf W (1994) 'Codesign from Cospecification', *IEEE Computer*, Vol. 21 , No. 1, 42–47.

Yourdon E (1989a) *Modern Structured Analysis*, Prentice-Hall, Englewood Cliffs, NJ, USA.

Yourdon E (1989b) *Structured Walkthroughs*, Prentice-Hall, Englewood Cliffs, NJ, USA.

Yourdon E and Constantine LL (1989) *Structured Design*, 2nd edition, Yourdon Press, Prentice-Hall, Englewood Cliffs, NJ, USA.

Zave P (1982) 'An Operational Approach to Requirements Specification for Embedded Systems', *IEEE Trans. on Soft. Eng.*, SE-9, No. 2, 250–269.

Zave P (1984) 'The Operational Versus the Conventional Approach to Software Development', *Comm. ACM*, Vol. 27, No. 2, 104–118.

SUBJECT INDEX